Dual-Number Methods in Kinematics, Statics and Dynamics

Ian S. Fischer

Dual-Number Methods in Kinematics, Statics and Dynamics

CRC Press
Boca Raton London New York Washington, D.C.

Library of Congress Cataloging-in-Publication Data

Fischer, Ian S.
Dual-number methods in kinematics, statics and dynamics / Ian S. Fischer.
p. cm.
Includes bibliographical references and index.
ISBN 0-8493-9115-6 (alk. paper)
1. Dynamics. 2. Mechanics, Analytic. 3. Kinematics. 4. Statics.
I. Title.
TA352.F57 1998
620.1′.04—dc21 98-34762
CIP

Direct all inquiries to CRC Press LLC, 2000 Corporate Blvd., N.W., Boca Raton, Florida 33431.

International Standard Book Number 0-8493-9115-6
Library of Congress Card Number 98-34762
Printed in the United States of America 1 2 3 4 5 6 7 8 9 0
Printed on acid-free paper

Preface

The use of the dual number allows compact formulations considerably facilitating the analysis of mechanical systems. While the methodology has been around for many years, it has remained obscure. The basic concepts of dual-number mechanics had been introduced by the beginning of the Twentieth Century, but it is only in recent decades that applications and developments have occasionally appeared in the technical literature.

In order to introduce students to dual-number methods, the author found it necessary to develop a course featuring application of dual numbers in the displacement, velocity, static and dynamic analysis of mechanisms. This course has been conducted on a more-or-less regular basis over the years, the lectures refined and has been popular with the students. While this course has introduced dual-number methodologies to the students who have taken it, it is realized that such a local effort does little to popularize dual-number methods amongst the academic community and design engineers in industry.

Consequently, the purpose of this volume is to popularize the dual-number analysis of mechanical systems. Certainly, dual-number methods have had little appeal because their explanations have been widely scattered about in the periodical literature, with a good many inconsistencies in manner of approach and in notation. What is needed is a book where a reasonably complete application of dual-number methods is discussed in a concise and consistent manner.

This book is designed to be used both as a text supporting a course using dual-number methods to analyze mechanisms and as a manual which can be used by an individual to develop abilities for the design of machinery or evaluation of mechanical systems. In addition to explanatory text and derivations, there are examples and exercises which can be used to gain insight and perfect skills.

The author notes that he was introduced to dual-number methods by his thesis advisor at Columbia University, Dr. Ferdinand Freudenstein. Much of what appears in the text was learned in conversations with his colleagues such as Dr. J. Michael McCarthy of the University of California, Irvine and Dr. Gordon R. Pennock of Purdue University. Comments on the manuscript by Dr. Pierre Larochelle of Florida Institute of Technology were very helpful. This book was written at New Jersey Institute of Technology and during a sabbatical spent at Pennsylvania State University hosted by Dr. Brian J. Gilmore, Dr. Donald A. Streit, Dr. Marc Carpino and their department head, Dr. Richard C. Benson. The American Society of Mechanical Engineers, the Society of Automotive Engineers, the Society of Applied Mechanisms and Robotics, Elsevier Science Ltd. and John Wiley & Sons have allowed use of illustrations from the author's works which they have published. Fundamental in facilitating the author's efforts was the support of the author's family, his parents Tobie B. Fischer and Clarence S. Fischer and brother Frederick S. Fischer.

Dr. Ian S. Fischer,
New Jersey Institute of Technology

Dedicated to the memory of Clarence S. Fischer

Table of Contents

Chapter 1 - Definitions

The Dual Number

The concept of the dual number was introduced by Clifford during 1873 and the name was given by Study in 1903.

The dual number ε is defined such that

$$\varepsilon \neq 0 \tag{1}$$

$$\varepsilon^2 = 0 \tag{2}$$

A dual number, which perhaps should be called a "duplex" number in analogy with complex numbers, is written as

$$\hat{a} = a + \varepsilon a_0 \tag{3}$$

where symbol a represents the primary (or real) part of duplex (or dual) number $\hat{a}$ and symbol a_0 represents the dual component of duplex (or dual) number $\hat{a}$.

In the mathematics of dual numbers, addition is defined as

$$\hat{a} + \hat{b} = (a + b) + \varepsilon(a_0 + b_0) \tag{4}$$

Multiplication is defined as

$$\hat{a}\hat{b} = (a + \varepsilon a_0)(b + \varepsilon b_0) \tag{5}$$

$$= ab + \varepsilon(ab_0 + a_0 b) \tag{6}$$

Division is handled in a manner similar to the division of complex numbers

$$\frac{\hat{a}}{\hat{b}} = \frac{a + \varepsilon a_0}{b + \varepsilon b_0} \tag{7}$$

$$= \frac{a + \varepsilon a_0}{b + \varepsilon b_0} \frac{b - \varepsilon b_0}{b - \varepsilon b_0} \tag{8}$$

$$= \frac{a}{b} + \varepsilon \frac{a_0 b - a b_0}{b^2} \tag{9}$$

Division by a pure dual number is not defined.

Raising a dual number to a power is

$$\hat{a}^n = (a + \varepsilon a_0)^n \tag{10}$$

$$= a^n + \varepsilon n a_0 a^{n-1} \tag{11}$$

Thus the square root of a dual number is

$$\sqrt{\hat{a}} = \sqrt{a} + \varepsilon \frac{a_0}{2\sqrt{a}} \tag{12}$$

The Taylor series expansion of a function of a dual number is

$$\mathrm{f}(\hat{a}) = \mathrm{f}(a + \varepsilon a_0) \tag{13}$$

$$= \mathrm{f}(a) + \varepsilon a_0 \frac{d\mathrm{f}(a)}{da} \tag{14}$$

All formal operations of dual number are the same as those of ordinary algebra followed by setting

$$\varepsilon^2 = \varepsilon^3 = \ldots = 0 \tag{15}$$

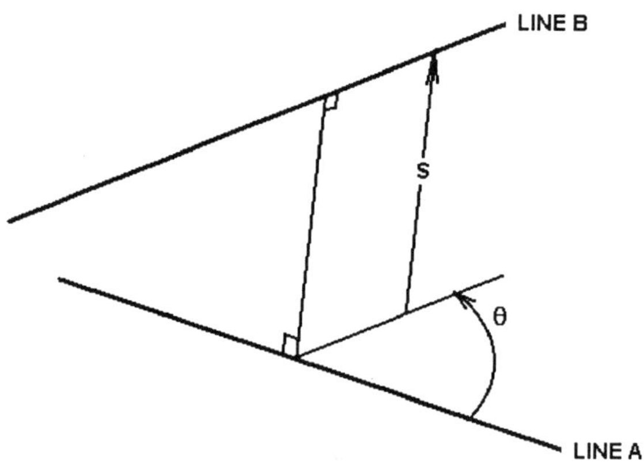

Figure 1 The dual angle $\hat{\theta} = \theta + \varepsilon s$ expresses the relationship between lines in space A and B.

The Dual Angle

The dual angle represents the relative displacement and orientation between two lines in space. This concept was introduced by Study (1903) and is illustrated in Figure 1. The dual angle is defined as

$$\hat{\theta} = \theta + \varepsilon s \tag{16}$$

where

θ = primary component of $\hat{\theta}$ = projected angle between lines A and B

s = dual component of $\hat{\theta}$ = shortest distance between lines A and B (length of common perpendicular).

If lines A and B intersect, dual angle $\hat{\theta}$ is a pure primary number. If the lines are parallel, dual angle $\hat{\theta}$ is a pure dual number.

Trigonometric Functions and Identities of Dual Angles

The trigonometric functions of a dual angle $\hat{\theta} = \theta + \varepsilon s$ are obtained using the Taylor-series expansion as specified in eq.(14). Consequently,

$$\sin\hat{\theta} = \sin\theta + \varepsilon s \cos\theta \tag{17}$$

$$\cos\hat{\theta} = \cos\theta - \varepsilon s \sin\theta \tag{18}$$

$$\tan\hat{\theta} = \tan\theta + \varepsilon s \sec^2\theta \tag{19}$$

Sometimes it will be more convenient to use the letters s and c instead of abbreviations sin and cos to represent the sine and cosine functions.

All identities of ordinary trigonometry hold true for dual angles. This axiom is known as the "Principle of Transference" or "Transfer Principle" and is attributed to Kotelnikov (1895) and Study (1901). Among the more useful trigonometric identities are

$$\sin^2\hat{\theta} + \cos^2\hat{\theta} = 1 \tag{20}$$

$$\sin 2\hat{\theta} = 2\sin\hat{\theta}\cos\hat{\theta} \tag{21}$$

$$\cos 2\hat{\theta} = \cos^2\hat{\theta} - \sin^2\hat{\theta} \tag{22}$$

$$\tan\left(\frac{\hat{\theta}}{2}\right) = \frac{1-\cos\hat{\theta}}{\sin\hat{\theta}} \tag{23}$$

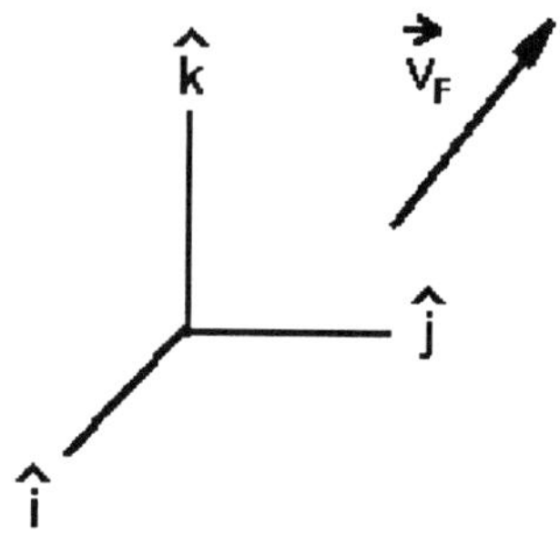

Figure 2 A free vector $\vec{V}_F$ must maintain its magnitude and direction but may be positioned anywhere in space.

Vectors

Free Vectors

A free vector has the qualities of magnitude and direction and, as shown in Figure 2, is represented pictorially as a line with an arrowhead whose length corresponds to its magnitude and whose orientation corresponds to its direction.

A free vector may be positioned anywhere in space so long as its magnitude and orientation are maintained.

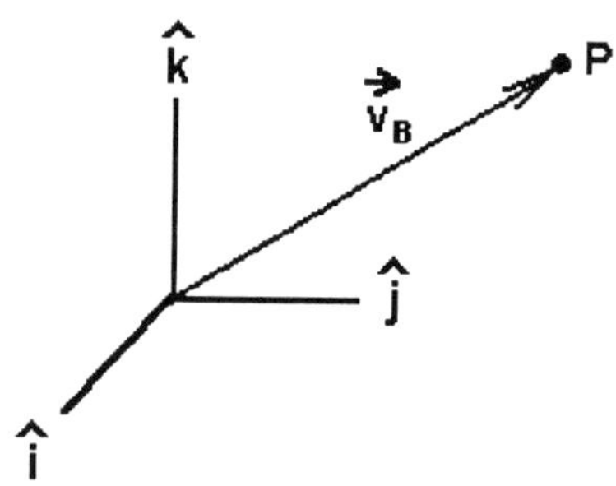

Figure 3 A bound vector $\vec{V}_B$ has its tail fixed to the origin of a coordinate system and is useful for locating a point *P*.

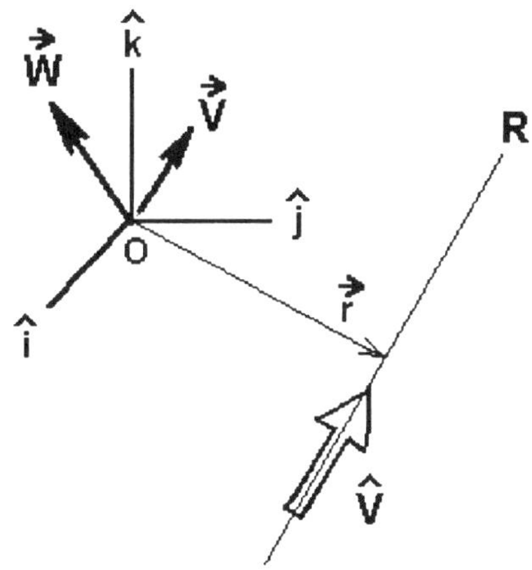

Figure 4 A dual vector $\hat{V} = \vec{V} + \varepsilon\,{}^{O}\vec{W} = \vec{V} + \varepsilon(\vec{r} \times \vec{V})$ has a given magnitude and is constrained to lie upon a given line *R* in space.

Bound Vectors

The tail of a bound vector is fixed at the origin of the coordinate system as is shown in Figure 3. It is used to specify the location of a point at its tip.

Dual Vectors

A vector constrained to lie upon a definite line in space, as shown in Figure 4, is called a "line vector". It is also called a "dual vector" when expressed in the form

$$\hat{V} = \vec{V} + \varepsilon\vec{W} \tag{24}$$

where the primary component $\vec{V}$ is called the "resultant vector" and comprises the magnitude and direction of dual vector $\hat{V}$. It is independent of the location of the coordinate system origin. The dual component $\vec{W}$ is called the "moment vector" and is defined as

$$\vec{W} = \vec{r} \times \vec{V} \tag{25}$$

where vector $\vec{r}$ connects origin *O* to any point on the line *R* of dual vector $\hat{V}$. Moment vector $\vec{W}$ is invariant for any choice of point on line *R*. However, the moment vector $\vec{W}$ does vary with the choice of coordinate system origin.

The dual vector can be expressed in terms of components along the coordinate-system unit vectors

$$\hat{\mathrm{V}} = V_x\vec{\mathrm{i}} + V_y\vec{\mathrm{j}} + V_z\vec{\mathrm{k}} + \varepsilon\left(W_x\vec{\mathrm{i}} + W_y\vec{\mathrm{j}} + W_z\vec{\mathrm{k}}\right) \tag{26}$$

The six scalar quantities comprising line vector $\hat{\mathrm{V}}$ are sometimes called "Plücker coordinates".

The Equation of Transformation

The "equation of transformation" allows expression of a dual vector given in terms of a coordinate system with origin at a point O in terms of another coordinate system similarly oriented with origin at another point P as is illustrated in Figure 5. A notation will be adopted in which a superscript on the left-hand side of a letter indicates the coordinate system in which the quantity is specified.

Now, consider a dual vector

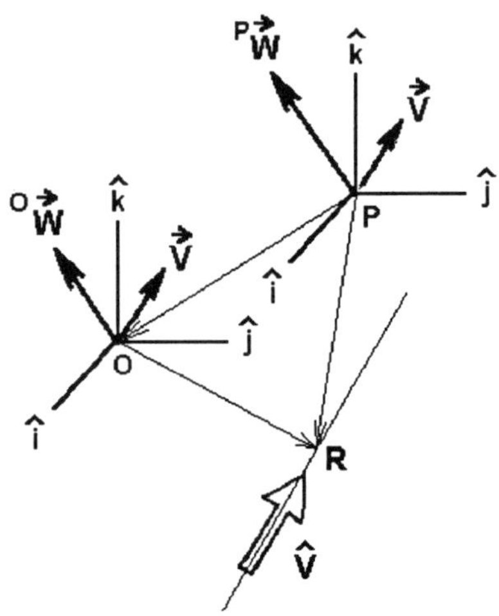

Figure 5 The equation of transformation allows a dual vector expressed in a frame {O} to be expressed in a similarly oriented frame {P}.

$$^{O}\hat{V} = \vec{V} + \varepsilon\, ^{O}\vec{W} \tag{27}$$

where resultant vector $\vec{V}$ is invariant with respect to origin and moment vector $\vec{W}$ varies with origin. Equation (27) can be rewritten as

$$^{O}\hat{V} = \vec{V} + \varepsilon\left(\vec{OR} \times \vec{V}\right) \tag{28}$$

where point R is any point on the line of dual vector $\hat{V}$.

Now, consider

$$^{P}\hat{V} = \vec{V} + \varepsilon\left(\vec{PR} \times \vec{V}\right) \tag{29}$$

where

$$\vec{PR} = \vec{PO} + \vec{OR} \tag{30}$$

So,

$$^{P}\hat{V} = \vec{V} + \varepsilon\left[\left(\vec{PO} + \vec{OR}\right) \times \vec{V}\right] \tag{31}$$

$$= \vec{V} + \varepsilon\left[\left(\vec{PO} \times \vec{V}\right) + \left(\vec{OR} \times \vec{V}\right)\right] \tag{32}$$

Recalling that

$$^{O}\vec{W} = \vec{OR} \times \vec{V} \tag{33}$$

So,

$$^{P}\hat{V} = \vec{V} + \varepsilon\left(^{O}\vec{W} + \vec{PO} \times \vec{V}\right) \tag{34}$$

This result is called the "equation of transformation" and transforms a dual vector $\hat{V}$ expressed in terms of the unit vectors of a coordinate frame {O} into terms of the unit vectors of a coordinate frame {P} which is oriented the same as frame {O} but whose origin is located by a translation through vector $\vec{OP}$ from the origin of frame {O}.

Exercises - Definitions

1. What is the sine of dual angle $37°-\varepsilon 7$?

2. Evaluate the sine of twice dual angle $15°-\varepsilon 3.2$, first by considering the doubled angle and then by using a trigonometric identity.

3. What is the expression for a dual vector which has magnitude 11.1803 directed along a unit vector $0.4472\vec{i}+0.5367\vec{j}+0.7155\vec{k}$ passing through a point with coordinates $\{2,4,9\}^T$?

4. A vector is expressed as $4\vec{i}+7\vec{j}-6\vec{k}$ in a coordinate frame through whose origin it passes. What is the expression for this same vector in a second coordinate frame oriented the same as the first frame and located at a point $13\vec{i}-52\vec{j}+14\vec{k}$ relative to the first frame?

Chapter 2 - Coordinate Transformation

Relative Displacements and Orientations

We wish to develop a mathematical technique to deal with the relative displacements and orientations of mechanism components. The technique should not only deal with displacements but be extendible to velocity and dynamic analysis.

Our scheme will be to define coordinate systems which are fixed on each link in the mechanism, and relate them through the mathematics of 3×3 dual-number coordinate transformations.

Screw Motion

The beginning of the screw-motion concept is attributed to Poinsot (1809) who established that any system of forces acting on a rigid body can be reduced to a single force and a single couple which are collinear. The idea was expanded when Chasles (1834) discovered that any rigid body displacement is equivalent to a simultaneous translation and rotation about a line in space. Other investigators examined the concept of screw motion, the most notable of which was R.S. Ball (1876).

Screw Motions With Respect to the Coordinate Axes

A coordinate frame (sometimes called a coordinate system) can be represented by unit vectors, vectors aligned with its axes and of unit length

$$ {}^{B}\hat{\mathrm{i}}_{A},\quad {}^{B}\hat{\mathrm{j}}_{A},\quad {}^{B}\hat{\mathrm{k}}_{A} $$

where the frame {A} specified by the unit vectors is indicated by the subscripts and the coordinate frame {B} in whose unit vectors these unit vectors are expressed is represented by the left superscripts.

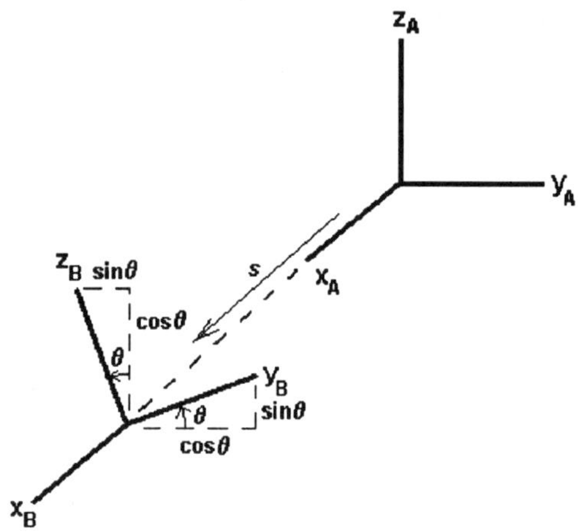

Figure 6 Coordinate frame {B} related to frame {A} by screw motion with respect to the *x*-axis of frame {A}.

Screw Motion With Respect to the X-axis

The unit vectors of a frame {A} can be written in the form of column vectors:

$$ {}^{A}\hat{\mathrm{i}}_{A} = \begin{Bmatrix} 1 \\ 0 \\ 0 \end{Bmatrix}, \quad {}^{A}\hat{\mathrm{j}}_{A} = \begin{Bmatrix} 0 \\ 1 \\ 0 \end{Bmatrix}, \quad {}^{A}\hat{\mathrm{k}}_{A} = \begin{Bmatrix} 0 \\ 0 \\ 1 \end{Bmatrix} \tag{35} $$

These column vectors can be combined into a single matrix describing frame {A}:

$$ \{\mathrm{A}\} = \left[{}^{A}\hat{\mathrm{i}}_{A},\ {}^{A}\hat{\mathrm{j}}_{A},\ {}^{A}\hat{\mathrm{k}}_{A} \right] \tag{36} $$

$$ \{\mathrm{A}\} = \begin{bmatrix} 1 & 0 & 0 \\ 0 & 1 & 0 \\ 0 & 0 & 1 \end{bmatrix} \tag{37} $$

Consider a screw motion through a dual angle $\hat{\theta}$ with respect to the x-axis of frame {A} as shown in Figure 6 where

$$ \hat{\theta} = \theta + \varepsilon s \tag{38} $$

in which angle θ represents the rotation about the screw axis and distance s represents the translation along the screw axis. The positive sense of the rotation will be defined by the "right-hand rule" as counter clockwise when looking into the positive end of the axis toward the origin.

The unit vectors of a frame {B} expressed in terms of the unit vectors of frame {A} are

$$ {}^{A}\hat{\mathrm{i}}_{B} = {}^{A}\hat{\mathrm{i}}_{A} = \begin{Bmatrix} 1 \\ 0 \\ 0 \end{Bmatrix} \tag{39} $$

and

$$ \begin{aligned} {}^{A}\hat{\mathrm{j}}_{B} &= \left(\cos\theta\,\vec{\mathrm{j}} + \sin\theta\,\vec{\mathrm{k}}\right) + \varepsilon\left[s\vec{\mathrm{i}} \times \left(\cos\theta\,\vec{\mathrm{j}} + \sin\theta\,\vec{\mathrm{k}}\right)\right] \\ &= \left(\cos\theta - \varepsilon s\sin\theta\right)\vec{\mathrm{j}} + \left(\sin\theta + \varepsilon s\cos\theta\right)\vec{\mathrm{k}} \\ &= \cos\hat{\theta}\,\vec{\mathrm{j}} + \sin\hat{\theta}\,\vec{\mathrm{k}} \end{aligned} $$

$$ {}^{A}\hat{\mathrm{j}}_{B} = \begin{Bmatrix} 0 \\ \cos\hat{\theta} \\ \sin\hat{\theta} \end{Bmatrix} \tag{40} $$

and

$$ \begin{aligned} {}^{A}\hat{\mathrm{k}}_{B} &= \left(-\sin\theta\,\vec{\mathrm{j}} + \cos\theta\,\vec{\mathrm{k}}\right) + \varepsilon\left[s\vec{\mathrm{i}} \times \left(-\sin\theta\,\vec{\mathrm{j}} + \cos\theta\,\vec{\mathrm{k}}\right)\right] \\ &= -\sin\hat{\theta}\,\vec{\mathrm{j}} + \cos\hat{\theta}\,\vec{\mathrm{k}} \end{aligned} $$

$$ {}^{A}\hat{\mathrm{k}}_{B} = \begin{Bmatrix} 0 \\ -\sin\hat{\theta} \\ \cos\hat{\theta} \end{Bmatrix} \tag{41} $$

Thus we can describe frame {B} in terms of the unit vectors of frame {A} as vectors ${}^{A}\hat{\mathrm{i}}_{B}$, ${}^{A}\hat{\mathrm{j}}_{B}$, ${}^{A}\hat{\mathrm{k}}_{B}$ combined into a single matrix

$$ \{\mathrm{B}\} = \left[{}^{A}\hat{\mathrm{i}}_{B},\ {}^{A}\hat{\mathrm{j}}_{B},\ {}^{A}\hat{\mathrm{k}}_{B}\right] \tag{42} $$

$$\{B\} = \begin{bmatrix} 1 & 0 & 0 \\ 0 & \cos\hat{\theta} & -\sin\hat{\theta} \\ 0 & \sin\hat{\theta} & \cos\hat{\theta} \end{bmatrix} \tag{43}$$

We seek an operator ${}^{A}_{B}\hat{X}$ which will map from frame {A} to frame {B} such that

$$\{B\} = {}^{A}_{B}\hat{X}\{A\} \tag{44}$$

The expressions for frames $\{A\}$ and $\{B\}$, eqs.(37,43) are substituted into eq.(44):

$$\begin{bmatrix} 1 & 0 & 0 \\ 0 & \cos\hat{\theta} & -\sin\hat{\theta} \\ 0 & \sin\hat{\theta} & \cos\hat{\theta} \end{bmatrix} = {}^{A}_{B}\hat{X} \begin{bmatrix} 1 & 0 & 0 \\ 0 & 1 & 0 \\ 0 & 0 & 1 \end{bmatrix} \tag{45}$$

This is expanded to show that

$${}^{A}_{B}\hat{X} = \begin{bmatrix} 1 & 0 & 0 \\ 0 & \cos\hat{\theta} & -\sin\hat{\theta} \\ 0 & \sin\hat{\theta} & \cos\hat{\theta} \end{bmatrix} \tag{46}$$

or

$${}^{A}_{B}\hat{X} = \left[{}^{A}\hat{i}_{B},\ {}^{A}\hat{j}_{B},\ {}^{A}\hat{k}_{B} \right] \tag{47}$$

The operator ${}^{A}_{B}\hat{X}$ is formally expressed as "the transformation matrix describing frame {B} relative to frame {A}." This operator is a description of the unit vectors of frame {B} in terms of the unit vectors of frame {A}.

The operator ${}^{A}_{B}\hat{X}$ allows any vector expressed in terms of the unit vectors of frame {B} to be expressed in terms of the unit vectors of frame {A}. Thus,

$${}^{A}\hat{V} = {}^{A}_{B}\hat{X}\,{}^{B}\hat{V} \tag{48}$$

where ${}^{A}\hat{V}$ is a dual vector expressed in frame {A} and ${}^{B}\hat{V}$ is the same dual vector expressed in frame {B}.

Let

$${}^{A}\hat{V} = {}^{A}\vec{V} + \varepsilon\,{}^{A}\vec{W} \tag{49}$$

$$ {}^{B}\hat{\mathrm{V}} = {}^{B}\vec{\mathrm{V}} + \varepsilon\, {}^{B}\vec{\mathrm{W}} \tag{50}$$

$$ {}^{A}_{B}\hat{\mathrm{X}} = {}^{A}_{B}\mathrm{R}_{X} + \varepsilon\, {}^{A}_{B}\mathrm{S}_{X} \tag{51}$$

It is observed that matrix ${}^{A}_{B}\mathrm{R}_{X}$ represents a simple rotation about the x-axis through angle θ and that matrix ${}^{A}_{B}\mathrm{S}_{X}$ represents a simple translation along the x-axis through distance s. So,

$$ {}^{A}\vec{\mathrm{V}} + \varepsilon\, {}^{A}\vec{\mathrm{W}} = \left({}^{A}_{B}\mathrm{R}_{X} + \varepsilon\, {}^{A}_{B}\mathrm{S}_{X}\right)\left({}^{B}\vec{\mathrm{V}} + \varepsilon\, {}^{B}\vec{\mathrm{W}}\right) \tag{52}$$

$$ = {}^{A}_{B}\mathrm{R}_{X}\, {}^{B}\vec{\mathrm{V}} + \varepsilon\left[{}^{A}_{B}\mathrm{R}_{X}\, {}^{B}\vec{\mathrm{W}} + {}^{A}_{B}S_{X}\, {}^{B}\vec{\mathrm{V}}\right] \tag{53}$$

Consider the primary component in eq.(53)

$$ {}^{A}\vec{\mathrm{V}} = {}^{A}_{B}\mathrm{R}_{X}\, {}^{B}\vec{\mathrm{V}} \tag{54}$$

The only change in the resultant vector is that it is described in terms of unit vectors oriented differently. (rotated through angle $-\theta$ about the x-axis).

Consider the dual component in eq.(53)

$$ {}^{B}\vec{\mathrm{W}} = {}^{A}_{B}\mathrm{R}_{X}\, {}^{B}\vec{\mathrm{W}} + \varepsilon\, {}^{A}_{B}\mathrm{S}_{X}\, {}^{B}\vec{\mathrm{V}} \tag{55}$$

The ${}^{A}_{B}\mathrm{R}_{X}\, {}^{B}\vec{\mathrm{W}}$ term represents moment vector ${}^{B}\vec{\mathrm{W}}$ described in terms of unit vectors oriented differently (rotated through angle $-\theta$ about the x-axis).

Let us examine the ${}^{A}_{B}\mathrm{S}_{X}\, {}^{B}\vec{\mathrm{V}}$ term. It can be expanded to

$$ {}^{A}_{B}\mathrm{S}_{X}\, {}^{B}\vec{\mathrm{V}} = s\begin{bmatrix} 0 & 0 & 0 \\ 0 & -\sin\theta & -\cos\theta \\ 0 & \cos\theta & -\sin\theta \end{bmatrix}\begin{Bmatrix} {}^{B}v_x \\ {}^{B}v_y \\ {}^{B}v_z \end{Bmatrix} \tag{56}$$

$$ = \begin{Bmatrix} 0 \\ s\left(-{}^{B}v_y \sin\theta - {}^{B}v_z \sin\theta\right) \\ s\left({}^{B}v_y \cos\theta - {}^{B}v_z \sin\theta\right) \end{Bmatrix} \tag{57}$$

It is observed that the right-hand side of this equation can be rewritten so that

$$ {}^{A}_{B}\mathrm{S}_{X}\, {}^{B}\vec{\mathrm{V}} = {}^{A}_{B}\mathrm{R}_{x}\left[-s\vec{\mathrm{i}} \times {}^{B}\vec{\mathrm{V}}\right] \tag{58}$$

This is equivalent to the second term of the moment vector in the equation of transformation expressed in unit vectors of frame $\{A\}$ (obtained by a screw motion through dual angle $-\hat{\theta} = -\theta - \varepsilon s$ with respect to the x-axis.

It is observed that the transformation-matrix method is compatible with the equation of transformation.

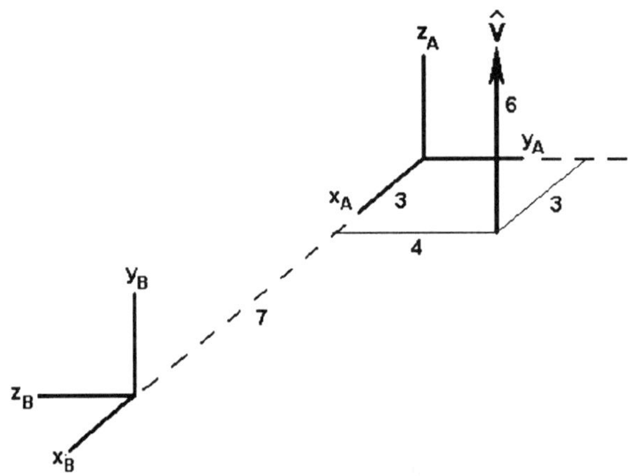

Figure 7 The vector $\hat{V}$ can be considered either in the unit vectors of frame {A} or those of frame {B}. Notice how the moment vector of $\hat{V}$ is $24\vec{i} - 18\vec{j} = (3\vec{i} + 4\vec{j}) \times 6\vec{k}$ with respect to frame {A} and $24\vec{i} - 42\vec{k} = (-7\vec{i} - 4\vec{k}) \times 6\vec{j}$ with respect to frame {B}.

Example

A frame {B} is arrived at by rotating 90° about and translating 10 units along the x-axis of frame {A} as shown in Figure 7. The transformation matrix stating the unit vectors of frame {B} in terms of the unit vectors of frame {A} is

$$ {}^A_B\hat{X} = \begin{bmatrix} 1 & 0 & 0 \\ 0 & \cos\hat{\theta} & -\sin\hat{\theta} \\ 0 & \sin\hat{\theta} & \cos\hat{\theta} \end{bmatrix} $$

where dual angle $\hat{\theta} = \frac{\pi}{2} + \varepsilon 10$ so that

$$
{}^{A}_{B}\hat{X} = \begin{bmatrix} 1 & 0 & 0 \\ 0 & -\varepsilon 10 & -1 \\ 0 & 1 & -\varepsilon 10 \end{bmatrix}
$$

Consider a line vector $\hat{V}$ described in frame {B} such that

$$
{}^{B}\hat{V} = \begin{Bmatrix} \varepsilon 24 \\ 6 \\ -\varepsilon 42 \end{Bmatrix}
$$

In frame {A} this vector is described as

$$
{}^{A}\hat{V} = {}^{A}_{B}\hat{X}\,{}^{B}\hat{V} = \begin{bmatrix} 1 & 0 & 0 \\ 0 & -\varepsilon 10 & -1 \\ 0 & 1 & -\varepsilon 10 \end{bmatrix} \begin{Bmatrix} \varepsilon 24 \\ 6 \\ -\varepsilon 42 \end{Bmatrix}
$$

$$
{}^{A}\hat{V} = \begin{Bmatrix} \varepsilon 24 \\ -\varepsilon 18 \\ 6 \end{Bmatrix}
$$

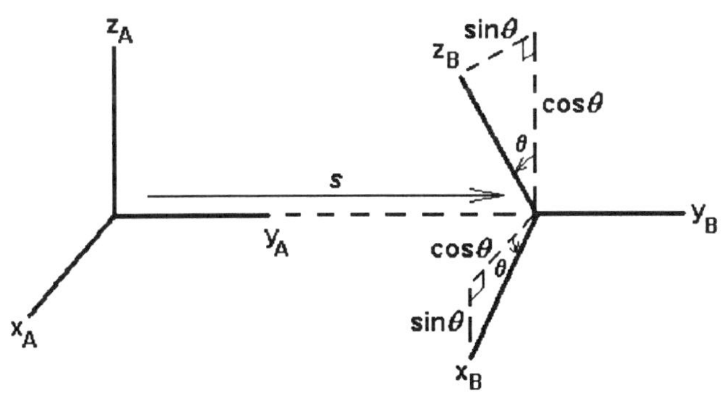

Figure 8 Coordinate frame {B} related to frame {A} by screw motion with respect to the *y*-axis of frame {A}.

Screw Motion With Respect to the Y-axis

Screw motion through a dual angle $\hat{\theta}$ about the *y*-axis, as illustrated in Figure 8, can be handled in a manner analogous to screw motion about the *x*-axis. Thus,

$$ {}^{A}_{B}\hat{Y} = \left[{}^{A}\hat{i}_B,\ {}^{A}\hat{j}_B,\ {}^{A}\hat{k}_B \right] \tag{59} $$

$$ {}^{A}_{B}\hat{Y} = \begin{bmatrix} \cos\hat{\theta} & 0 & \sin\hat{\theta} \\ 0 & 1 & 0 \\ -\sin\hat{\theta} & 0 & \cos\hat{\theta} \end{bmatrix} \tag{60} $$

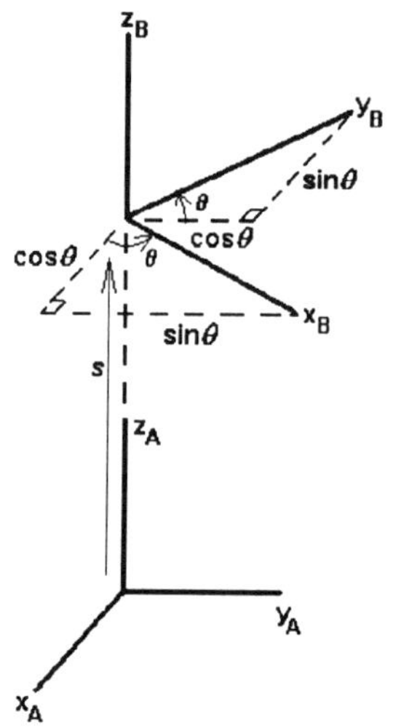

Figure 9 Coordinate frame {B} related to frame {A} by screw motion with respect to the *z*-axis of frame {A}.

Screw Motion With Respect to the Z-axis

Screw motion through a dual angle $\hat{\theta}$ about the *z*-axis, shown in Figure 9, is also dealt with in the same way as screw motion about either of the *x*-axis or *y*-axis. Thus,

$$ {}^{A}_{B}\hat{Z} = \left[{}^{A}\hat{i}_B,\ {}^{A}\hat{j}_B,\ {}^{A}\hat{k}_B \right] \tag{61} $$

$$ {}^{A}_{B}\hat{Z} = \begin{bmatrix} \cos\hat{\theta} & -\sin\hat{\theta} & 0 \\ \sin\hat{\theta} & \cos\hat{\theta} & 0 \\ 0 & 0 & 1 \end{bmatrix} \tag{62} $$

Motion of a Vector Through a Dual Angle

We have considered the situation where the vector is stationary and the new frame is moved relative to the old frame. We can apply kinematic inversion and consider the moving of the vector relative to a stationary frame.

Imagine that a vector moves from position 1 to position 2 while the frame remains stationary. If an observer was fixed on the vector, and the motion was conducted so smoothly that it was imperceptible to the observer, he would think that the frame was moving. The appearance is that the frame moves in directions reverse from the actual motion of the vector relative to the fixed frame.

$$\hat{V}_2 = {}^2_1\hat{T}\hat{V}_1 \tag{63}$$

Thus the same mathematics dealing with moving frames and stationary vectors can be used for stationary frames and moving vectors. It is observed that vector $\hat{V}_1$ and vector $\hat{V}_2$ are different vectors as suggested by their different subscripts, but that they are both expressed in the unit vectors of the same frame.

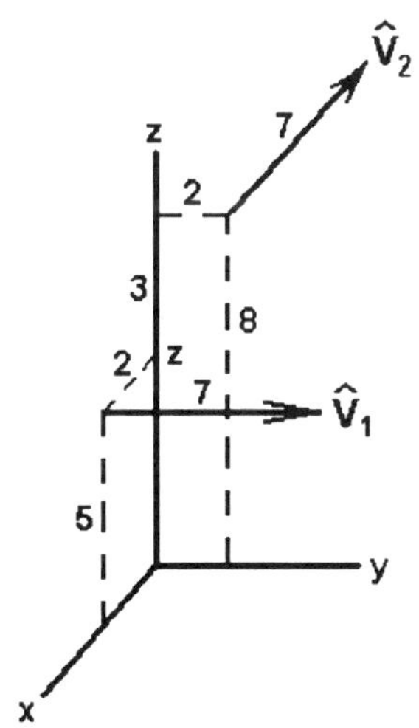

Figure 10 Example of the screw motion of a vector with respect to the *z*-axis.

Example

Consider a vector

$$\hat{V} = \begin{Bmatrix} -\varepsilon 35 \\ 7 \\ \varepsilon 14 \end{Bmatrix}$$

Effect a rotation of this vector through dual angle $\hat{\theta} = 90° + \varepsilon 3$ about the z-axis as shown in Figure 10.

$$\hat{V}_2 = {}^2_1\hat{Z}\hat{V}_1$$

$$= \begin{bmatrix} -\varepsilon 3 & -1 & 0 \\ 1 & -\varepsilon 3 & 0 \\ 0 & 0 & 1 \end{bmatrix} \begin{Bmatrix} -\varepsilon 3 \\ 7 \\ \varepsilon 14 \end{Bmatrix}$$

$$\hat{V}_2 = \begin{Bmatrix} -7 \\ -\varepsilon 56 \\ \varepsilon 14 \end{Bmatrix}$$

Inverse Transformation

There exists a unique 3×3 matrix $\hat{I}$ called the "dual identity matrix" such that

$$\hat{M} = \hat{I}\hat{M} \tag{64}$$

where $\hat{M}$ is any vector or matrix for which the multiplication can be performed.

Now,

$$\hat{I} = \begin{bmatrix} 1 & 0 & 0 \\ 0 & 1 & 0 \\ 0 & 0 & 1 \end{bmatrix} \tag{65}$$

$$= \begin{bmatrix} 1 & 0 & 0 \\ 0 & 1 & 0 \\ 0 & 0 & 1 \end{bmatrix} + \varepsilon \begin{bmatrix} 0 & 0 & 0 \\ 0 & 0 & 0 \\ 0 & 0 & 0 \end{bmatrix} \tag{66}$$

$$= I + \varepsilon[0] \tag{67}$$

where symbol I represents the 3×3 identity matrix and symbol [0] represents the 3×3 null matrix.

Because operating the dual identity matrix on a vector or frame produces no change, the action is equivalent to remaining stationary.

Consider a transformation

$$ {}^{2}\hat{V} = {}^{2}_{1}\hat{T}\, {}^{1}\hat{V} \tag{68} $$

and another transformation identical in magnitudes but opposite in directions

$$ {}^{1}\hat{V} = {}^{1}_{2}\hat{T}\, {}^{2}\hat{V} \tag{69} $$

Substitute eq.(68) into eq.(69),

$$ {}^{1}\hat{V} = {}^{1}_{2}\hat{T}\, {}^{2}_{1}\hat{T}\, {}^{1}\hat{V} \tag{70} $$

Canceling the common terms,

$$ \hat{I} = {}^{1}_{2}\hat{T}\, {}^{2}_{1}\hat{T} \tag{71} $$

This is apparent since the ${}^{1}_{2}\hat{T}\, {}^{2}_{1}\hat{T}$ operator produces no change on dual vector ${}^{1}\hat{V}$.

If the product of two matrices produces the dual identity matrix $\hat{I}$, then one matrix is the "inverse" of the other.

This is written as

$$ {}^{1}_{2}\hat{T} = {}^{2}_{1}\hat{T}^{-1} \tag{72} $$

or

$$ {}^{2}_{1}\hat{T} = {}^{1}_{2}\hat{T}^{-1} \tag{73} $$

Now, it is observed that the value of the determinant of any dual-number coordinate-transformation matrix is unity. This implies that a dual-number coordinate-transformation matrix is "orthogonal." For an orthogonal matrix, the inverse equals the transpose.

$$ {}^{1}_{2}\hat{T} = {}^{2}_{1}\hat{T}^{-1} = {}^{2}_{1}\hat{T}^{T} \tag{74} $$

Thus, to obtain the inverse of a dual-number coordinate-transformation matrix, we need only interchange the order of the subscripts of the elements. This is simple computationally and a big advantage over homogeneous coordinate-transformation matrices where finding the inverse involves many arithmetic operations.

Example

The inverse of

$$\begin{bmatrix} \varepsilon 12 & 0 & 1 \\ 0 & 1 & 0 \\ -1 & 0 & \varepsilon 12 \end{bmatrix}$$

is

$$\begin{bmatrix} \varepsilon 12 & 0 & -1 \\ 0 & 1 & 0 \\ 1 & 0 & \varepsilon 12 \end{bmatrix}$$

Proof:

$$\begin{bmatrix} \varepsilon 12 & 0 & 1 \\ 0 & 1 & 0 \\ -1 & 0 & \varepsilon 12 \end{bmatrix}\begin{bmatrix} \varepsilon 12 & 0 & -1 \\ 0 & 1 & 0 \\ 1 & 0 & \varepsilon 12 \end{bmatrix} = \begin{bmatrix} 1 & 0 & 0 \\ 0 & 1 & 0 \\ 0 & 0 & 1 \end{bmatrix}$$

Successive Transformations

Successive transformations about fixed coordinate axes are handled by successive left multiplication.

Consider a screw motion from {1} to {2} followed by a screw motion from {2} to {3} and then by a screw motion from {3} to {4}. Let the transformation matrices be ${}^{2}_{1}\hat{T}$, ${}^{3}_{2}\hat{T}$, ${}^{4}_{3}\hat{T}$.

To make the individual transformations, we have

$$ {}^{2}\hat{V} = {}^{2}_{1}\hat{T}\,{}^{1}\hat{V} \tag{75}$$

$$ {}^{3}\hat{V} = {}^{3}_{2}\hat{T}\,{}^{2}\hat{V} \tag{76}$$

$$ {}^{4}\hat{V} = {}^{4}_{3}\hat{T}\,{}^{3}\hat{V} \tag{77}$$

Substituting eq.(76) into eq.(77)

$$ {}^{4}\hat{V} = {}^{4}_{3}\hat{T}\,{}^{3}_{2}\hat{T}\,{}^{2}\hat{V} \tag{78}$$

and eq.(75) into eq.(78)

$$ {}^{4}\hat{V} = {}^{4}_{3}\hat{T}\,{}^{3}_{2}\hat{T}\,{}^{2}_{1}\hat{T}\,{}^{1}\hat{V} \tag{79}$$

Thus the successive transformations can be concatenated into a single transformation so that

$$ {}^{4}_{1}\hat{T} = {}^{4}_{3}\hat{T}\,{}^{3}_{2}\hat{T}\,{}^{2}_{1}\hat{T} \tag{80} $$

and

$$ {}^{4}\hat{V} = {}^{4}_{1}\hat{T}\,{}^{1}\hat{V} \tag{81} $$

This concept works for any number of transformations. Generalized, the concatenation of transformations takes the form

$$ {}^{Z}_{A}T = {}^{Z}_{Z-1}\hat{T}\,{}^{Z-1}_{Z-2}\hat{T}\ldots{}^{A+2}_{A+1}\hat{T}\,{}^{A+1}_{A}\hat{T} \tag{82} $$

It is noted that the order of the screw motions is important.

Also, it is noted that in eq.(82) that all the motions are with respect to frame {A}, the fixed frame, regardless of the intermediate positions and orientations of the moving vector or frame.

Example

Make a screw motion through dual angle $\hat{\theta}_1 = 30°+\varepsilon 4$ about the *x*-axis of the fixed frame followed by a screw motion through dual angle $\hat{\theta}_2 = 20°+\varepsilon 3$ about the *y*-axis of the same fixed frame.

$$ {}^{A}_{B}\hat{T}_{xy} = \begin{bmatrix} c\hat{\theta}_2 & 0 & s\hat{\theta}_2 \\ 0 & 1 & 0 \\ -s\hat{\theta}_2 & 0 & c\hat{\theta}_2 \end{bmatrix}\begin{bmatrix} 1 & 0 & 0 \\ 0 & c\hat{\theta}_1 & -s\hat{\theta}_1 \\ 0 & s\hat{\theta}_1 & c\hat{\theta}_1 \end{bmatrix} = \begin{bmatrix} c\hat{\theta}_2 & s\hat{\theta}_1 s\hat{\theta}_2 & c\hat{\theta}_1 s\hat{\theta}_2 \\ 0 & c\hat{\theta}_1 & -s\hat{\theta}_1 \\ -s\hat{\theta}_2 & s\hat{\theta}_1 c\hat{\theta}_2 & c\hat{\theta}_1 c\hat{\theta}_2 \end{bmatrix} $$

where letters s and c are used to represent the sine and cosine functions.

Upon substituting the values of the dual angles, we obtain

$$ {}^{A}_{B}\hat{T}_{xy} = \begin{bmatrix} 0.9397-\varepsilon 1.026 & 0.171+\varepsilon 2.594 & 0.2962+\varepsilon 1.757 \\ 0 & 0.866-\varepsilon 2 & -0.5-\varepsilon 3.464 \\ -0.342-\varepsilon 2.082 & 0.4698+\varepsilon 2.742 & 0.8138-\varepsilon 2.768 \end{bmatrix} $$

Now consider the same two screw motions but in reverse order.

$$ {}^{A}_{B}\hat{T}_{yx} = \begin{bmatrix} 1 & 0 & 0 \\ 0 & c\hat{\theta}_1 & -s\hat{\theta}_1 \\ 0 & s\hat{\theta}_1 & c\hat{\theta}_1 \end{bmatrix}\begin{bmatrix} c\hat{\theta}_2 & 0 & s\hat{\theta}_2 \\ 0 & 1 & 0 \\ -s\hat{\theta}_2 & 0 & c\hat{\theta}_2 \end{bmatrix} = \begin{bmatrix} c\hat{\theta}_2 & 0 & s\hat{\theta}_2 \\ s\hat{\theta}_1 s\hat{\theta}_2 & c\hat{\theta}_1 & -s\hat{\theta}_1 c\hat{\theta}_2 \\ -c\hat{\theta}_1 s\hat{\theta}_2 & s\hat{\theta}_1 & c\hat{\theta}_1 c\hat{\theta}_2 \end{bmatrix} $$

Note how this differs from the previous result.

Upon substituting the values of the dual angles, we obtain

$$ {}^{A}_{B}\hat{T}_{yx} = \begin{bmatrix} 0.9397-\varepsilon 1.026 & 0 & 0.342+\varepsilon 2.819 \\ 0.171+\varepsilon 2.594 & 0.8667-\varepsilon 2 & -0.4698-\varepsilon 2.742 \\ -0.2962-\varepsilon 1.757 & 0.5+\varepsilon 3.464 & 0.8138-\varepsilon 2.768 \end{bmatrix} $$

Roll-Pitch-Yaw About the Fixed Axes

Consider a coordinate frame fixed in space. Roll through a dual angle is a screw motion with respect to the x-axis of that frame. Pitch through a dual angle is a screw motion with respect to the y-axis, and yaw through a dual angle is a screw motion with respect to the z-axis of the coordinate frame.

Roll($\hat{\gamma}$) = Screw motion $\hat{\gamma} = \gamma + \varepsilon G$ about the x-axis

Pitch($\hat{\beta}$) = Screw motion $\hat{\beta} = \beta + \varepsilon B$ about the y-axis

Yaw($\hat{\alpha}$) = Screw motion $\hat{\alpha} = \alpha + \varepsilon A$ about the z-axis

A combination of these is

$$ {}^{A}_{B}\hat{T}_{RPY}\left(\hat{\gamma},\hat{\beta},\hat{\alpha}\right) = \mathrm{Roll}(\hat{\alpha})\,\mathrm{Pitch}(\hat{\beta})\,\mathrm{Yaw}(\hat{\gamma}) = \hat{Z}(\hat{\alpha})\,\hat{Y}(\hat{\beta})\,\hat{X}(\hat{\gamma}) \tag{83} $$

$$ = \begin{bmatrix} c\hat{\alpha} & -s\hat{\alpha} & 0 \\ s\hat{\alpha} & c\hat{\alpha} & 0 \\ 0 & 0 & 1 \end{bmatrix} \begin{bmatrix} c\hat{\beta} & 0 & s\hat{\beta} \\ 0 & 1 & 0 \\ -s\hat{\beta} & 0 & c\hat{\beta} \end{bmatrix} \begin{bmatrix} 1 & 0 & 0 \\ 0 & c\hat{\gamma} & -s\hat{\gamma} \\ 0 & s\hat{\gamma} & c\hat{\gamma} \end{bmatrix} \tag{84} $$

$$ {}^{A}_{B}\hat{T}_{RPY}\left(\hat{\gamma},\hat{\beta},\hat{\alpha}\right) = \begin{bmatrix} c\hat{\alpha}c\hat{\beta} & c\hat{\alpha}s\hat{\beta}s\hat{\gamma} - s\hat{\alpha}c\hat{\gamma} & c\hat{\alpha}s\hat{\beta}c\hat{\gamma} + s\hat{\alpha}s\hat{\gamma} \\ s\hat{\alpha}c\hat{\beta} & s\hat{\alpha}s\hat{\beta}s\hat{\gamma} + c\hat{\alpha}c\hat{\gamma} & c\hat{\alpha}s\hat{\beta}s\hat{\gamma} - c\hat{\alpha}s\hat{\gamma} \\ s\hat{\beta} & c\hat{\beta}s\hat{\gamma} & c\hat{\beta}c\hat{\gamma} \end{bmatrix} \tag{85} $$

$$
{}^{A}_{B}\hat{T}_{RPY} =
\begin{bmatrix}
\begin{array}{c} c\alpha c\beta \\ -\varepsilon\begin{bmatrix} A s\alpha c\beta \\ +B c\alpha s\beta \end{bmatrix} \end{array} &
\begin{array}{c} -s\alpha c\gamma + c\alpha s\beta s\gamma \\ +\varepsilon\begin{bmatrix} -A(c\alpha c\gamma + s\alpha s\beta s\gamma) \\ +B c\alpha c\beta s\gamma \\ -G(s\alpha s\gamma + c\alpha s\beta c\gamma) \end{bmatrix} \end{array} &
\begin{array}{c} s\alpha s\gamma + c\alpha s\beta c\gamma \\ +\varepsilon\begin{bmatrix} A(c\alpha s\gamma - s\alpha s\beta c\gamma) \\ +B c\alpha c\beta c\gamma \\ +G(s\alpha c\gamma - c\alpha s\beta s\gamma) \end{bmatrix} \end{array} \\
\begin{array}{c} s\alpha c\beta \\ +\varepsilon\begin{bmatrix} A c\alpha c\beta \\ -B s\alpha s\beta \end{bmatrix} \end{array} &
\begin{array}{c} c\alpha c\gamma + s\alpha s\beta s\gamma \\ +\varepsilon\begin{bmatrix} A(-s\alpha c\gamma + c\alpha s\beta s\gamma) \\ +B s\alpha c\beta s\gamma \\ +G(-c\alpha s\gamma + s\alpha s\beta c\gamma) \end{bmatrix} \end{array} &
\begin{array}{c} -c\alpha s\gamma + s\alpha s\beta c\gamma \\ +\varepsilon\begin{bmatrix} A(s\alpha s\gamma + c\alpha s\beta c\gamma) \\ +B s\alpha c\beta c\gamma \\ -G(c\alpha c\gamma + s\alpha s\beta s\gamma) \end{bmatrix} \end{array} \\
\begin{array}{c} -s\beta \\ -\varepsilon B c\beta \end{array} &
\begin{array}{c} c\beta s\gamma \\ +\varepsilon\begin{bmatrix} -B s\beta s\gamma \\ +G c\beta c\gamma \end{bmatrix} \end{array} &
\begin{array}{c} c\beta c\gamma \\ +\varepsilon\begin{bmatrix} -B s\beta c\gamma \\ -G c\beta s\gamma \end{bmatrix} \end{array}
\end{bmatrix} \quad (86)
$$

It is important to note that the order of the motions is important. Sequencing the roll, pitch and yaw differently will result in an expression different than eq.(86).

Example

Evaluate the dual angles of roll, pitch and yaw (assuming that order) for the coordinate-transformation matrix

$$
{}^{A}_{B}\hat{T}_{RPY} = \begin{bmatrix} 0.9254 - \varepsilon 1.337 & 0.01806 + \varepsilon 1.137 & 0.3785 + \varepsilon 3.214 \\ 0.1631 + \varepsilon 1.673 & 0.8825 - \varepsilon 1.483 & -0.4410 - \varepsilon 2.349 \\ -0.3420 - \varepsilon 2.819 & 0.4698 + \varepsilon 2.7421 & 0.8138 - \varepsilon 2.768 \end{bmatrix}
$$

Designating the primary and dual components of element $\hat{t}_{ij}$ as $\hat{t}_{ijp}$ and $\hat{t}_{ijd}$ respectively,

$$
\tan\gamma = \frac{t_{32p}}{t_{33p}} = \frac{c\beta s\gamma}{c\beta c\gamma} = \frac{0.4698}{0.8138} = 0.5774
$$

$$
\tan^{-1} 0.5774 = 30°, 210°
$$

In that the numerator and denominator in the expression for $\tan\gamma$ are both positive, the angle must reside in the first quadrant, and so

$$
\gamma = 30°
$$

Now,

$$s\beta = -t_{31p} = -(-s\beta) = 0.3420$$

$$c\beta = \sqrt{t_{11p}^2 + t_{21p}^2} = \sqrt{(c\alpha c\beta)^2 + (s\alpha c\beta)^2} = \sqrt{(0.9254)^2 + (0.1631)^2} = 0.9397$$

so

$$\tan\beta = \frac{s\beta}{c\beta} = \frac{0.3420}{0.9397} = 0.3639$$

which indicates that

$$\beta = 20°$$

Now,

$$\tan\alpha = \frac{t_{21p}}{t_{11p}} = \frac{0.1631}{0.9254} = 0.1763$$

so that

$$\alpha = 10°$$

To compute the translations,

$$B = \frac{-t_{31d}}{c\beta} = \frac{-(-Bc\beta)}{c\beta} = \frac{-(-2.819)}{0.9256} = 3$$

$$A = \frac{t_{21d} + Bs\alpha\, s\beta}{c\alpha c\beta} = \frac{1.673 + 3(0.1736)(0.3420)}{(0.9848)(0.9397)} = 2$$

$$G = \frac{t_{32d} + Bs\beta\, s\gamma}{c\beta c\gamma} = \frac{2.7421 + 3(0.3420)(0.5)}{(0.9397)(0.8660)} = 4$$

ZYX Euler Angles

Consider an object moving through space. In the preceding discussion, we considered screw motions with respect to a coordinate system which was fixed in space. When the body went through a series of motions, each motion was referred to the axes of that same coordinate frame.

We can also express the motion of the body with respect to the axes of a coordinate frame fixed on the body. When each screw motion has been completed, the location and orientation of that coordinate frame has changed since the body on which it is fixed has moved. Each successive screw motion occurs about an axis whose location and orientation is dependent on preceding motions. We say that the screw motions are taken about the axes of a moving coordinate system. The dual angles expressing the screw motions about the axes of a moving coordinate frame are called Euler dual angles.

We can consider screw motions through Euler dual angles about any sequence of axes as long as we do not repeat the same axis consecutively. Let us consider the ZYX Euler dual angles where the screw motions occur about the axes $\vec{X},\vec{Y},\vec{Z}$. These will be referred to as ZYX Euler dual angles. It is noted that ZYZ Euler dual angles are another popular form.

Consider a frame {A} in space which will undergo screw motions through the ZYX Euler dual angles $\hat{\alpha},\hat{\beta},\hat{\gamma}$ to move into alignment with a frame {B}. The motion will occur in three steps:

Rotate frame {A} about axis $\hat{z}_A$ through angle $\hat{\alpha}$ to frame {B′}.

Rotate frame {B′} about axis $\hat{y}_{B'}$ through angle $\hat{\beta}$ to frame {B″}.

Rotate frame {B″} about axis $\hat{x}_{B''}$ through angle $\hat{\gamma}$ to frame {B}.

We now wish to evaluate a coordinate-transformation matrix ${}^A_B E(\hat{\alpha},\hat{\beta},\hat{\gamma})$ which will transform a coordinate frame {A} through Euler dual angles $\hat{\alpha},\hat{\beta},\hat{\gamma}$ into alignment with coordinate frame {B}.

The problem is that the three component screw motions are expressed in different frames, and the previously developed method of concatenating successive transforms deals only with screw motions referred to the axis of the same frame. To combine screw motions about the axes of three different frames into a single transformation matrix, the motions must all be referred to the same set of axes.

Clearly, our definition of Euler dual angles consists of screw motions with respect to axes of different coordinate frames. Thus, our formulation of the motions through Euler angles calls for a different approach than our prior consideration of successive transformations with respect to a fixed frame. Essentially we have a coordinate system fixed on the moving body and a coordinate system fixed in the world. According to our definitions, we are taking our screw motions with respect to the system on the moving body.

Consider a person riding the body as it moves through space. If the motion were smooth enough that the accelerations could not be felt, the person would feel that he was still and the world was moving. Furthermore, he would sense that the world was moving in a direction opposite of the actual motion of the body. For instance, the passenger on a westbound train feels that he is stationary and the scenery seen through a window is moving east.

So, let us consider the part stationary and move the world. That way we can take all our screw motions with respect to the same frame.

Instead of looking at the first motion as a screw motion with respect to the z_A-axis (world system) through dual angle $\hat{\alpha}$, consider it as a screw motion with respect to the $z_{B'}$-axis through dual angle $-\hat{\alpha}$.

Instead of looking at the second motion as a screw motion with respect to the $y_{B'}$-axis through dual angle $\hat{\beta}$, consider it as a screw motion with respect to the $y_{B''}$-axis through dual angle $-\hat{\beta}$.

Instead of looking at the third motion as a screw motion with respect to the $x_{B''}$-axis through dual angle $\hat{\gamma}$, consider it as a screw motion with respect to the $x_{B'''}$-axis through dual angle $-\hat{\gamma}$.

So,

$$ {}^{B}_{A}\hat{E} = \hat{X}(-\hat{\gamma})\hat{Y}(-\hat{\beta})\hat{Z}(-\hat{\alpha}) \tag{87} $$

Now,

$$ \hat{X}(-\hat{\gamma}) = \hat{X}^{-1}(\hat{\gamma}) \tag{88} $$

$$ \hat{Y}(-\hat{\beta}) = \hat{Y}^{-1}(\hat{\beta}) \tag{89} $$

$$ \hat{Z}(-\hat{\alpha}) = \hat{Z}^{-1}(\hat{\alpha}) \tag{90} $$

So,

$$ {}^{B}_{A}\hat{E} = \hat{X}^{-1}(\hat{\gamma})\hat{Y}^{-1}(\hat{\beta})Z^{-1}(\hat{\alpha}) \tag{91} $$

$$ {}^{B}_{A}\hat{E}\,Z(\hat{\alpha}) = \hat{X}^{-1}(\hat{\gamma})\,\hat{Y}^{-1}(\hat{\beta}) \tag{92} $$

$$ {}^{B}_{A}\hat{E}\,Z(\hat{\alpha})\,\hat{Y}(\hat{\beta}) = \hat{X}^{-1}(\hat{\gamma}) \tag{93} $$

$$ {}^{B}_{A}\hat{E}\,Z(\hat{\alpha})\hat{Y}(\hat{\beta})\hat{X}(\hat{\gamma}) = \hat{I} \tag{94} $$

$$ Z(\hat{\alpha})\,\hat{Y}(\hat{\beta})\,\hat{X}(\hat{\gamma}) = {}^{B}_{A}\hat{E}^{-1} \tag{95} $$

Now,

$$ {}^{A}_{B}\hat{\mathrm{E}} = {}^{B}_{A}\hat{\mathrm{E}}^{-1} \tag{96} $$

So,

$$ {}^{A}_{B}\hat{\mathrm{E}} = \mathrm{Z}(\hat{\alpha})\,\hat{\mathrm{Y}}(\hat{\beta})\,\hat{\mathrm{X}}(\hat{\gamma}) \tag{97} $$

$$ = \begin{bmatrix} c\hat{\alpha} & -s\hat{\alpha} & 0 \\ s\hat{\alpha} & c\hat{\alpha} & 0 \\ 0 & 0 & 1 \end{bmatrix} \begin{bmatrix} c\hat{\beta} & 0 & s\hat{\beta} \\ 0 & 1 & 0 \\ -s\hat{\beta} & 0 & c\hat{\beta} \end{bmatrix} \begin{bmatrix} 1 & 0 & 0 \\ 0 & c\hat{\gamma} & -s\hat{\gamma} \\ 0 & s\hat{\gamma} & c\hat{\gamma} \end{bmatrix} \tag{98} $$

$$ {}^{A}_{B}\hat{\mathrm{E}} = \begin{bmatrix} c\hat{\alpha}c\hat{\beta} & c\hat{\alpha}s\hat{\beta}s\hat{\gamma} - s\hat{\alpha}c\hat{\gamma} & c\hat{\alpha}s\hat{\beta}c\hat{\gamma} + s\hat{\alpha}s\hat{\gamma} \\ s\hat{\alpha}c\hat{\beta} & s\hat{\alpha}s\hat{\beta}s\hat{\gamma} + c\hat{\alpha}c\hat{\gamma} & c\hat{\alpha}s\hat{\beta}s\hat{\gamma} - c\hat{\alpha}s\hat{\gamma} \\ s\hat{\beta} & c\hat{\beta}s\hat{\gamma} & c\hat{\beta}c\hat{\gamma} \end{bmatrix} \tag{99} $$

$$ {}^{A}_{B}\hat{\mathrm{E}} = \begin{bmatrix} \begin{matrix} c\alpha c\beta \\ -\varepsilon\begin{bmatrix} A s\alpha c\beta \\ + B c\alpha s\beta \end{bmatrix} \end{matrix} & \begin{matrix} -s\alpha c\gamma + c\alpha s\beta s\gamma \\ +\varepsilon\begin{bmatrix} -A(c\alpha c\gamma + s\alpha s\beta s\gamma) \\ + B c\alpha c\beta s\gamma \\ - G(s\alpha s\gamma + c\alpha s\beta c\gamma) \end{bmatrix} \end{matrix} & \begin{matrix} s\alpha s\gamma + c\alpha s\beta c\gamma \\ +\varepsilon\begin{bmatrix} A(c\alpha s\gamma - s\alpha s\beta c\gamma) \\ + B c\alpha c\beta c\gamma \\ + G(s\alpha c\gamma - c\alpha s\beta s\gamma) \end{bmatrix} \end{matrix} \\ \begin{matrix} s\alpha c\beta \\ +\varepsilon\begin{bmatrix} A c\alpha c\beta \\ - B s\alpha s\beta \end{bmatrix} \end{matrix} & \begin{matrix} c\alpha c\gamma + s\alpha s\beta s\gamma \\ +\varepsilon\begin{bmatrix} A(-s\alpha c\gamma + c\alpha s\beta s\gamma) \\ + B s\alpha c\beta s\gamma \\ + G(-c\alpha s\gamma + s\alpha s\beta c\gamma) \end{bmatrix} \end{matrix} & \begin{matrix} -c\alpha s\gamma + s\alpha s\beta c\gamma \\ +\varepsilon\begin{bmatrix} A(s\alpha s\gamma + c\alpha s\beta c\gamma) \\ + B s\alpha c\beta c\gamma \\ - G(c\alpha c\gamma + s\alpha s\beta s\gamma) \end{bmatrix} \end{matrix} \\ \begin{matrix} -s\beta \\ -\varepsilon B c\beta \end{matrix} & \begin{matrix} c\beta s\gamma \\ +\varepsilon\begin{bmatrix} -B s\beta s\gamma \\ + G c\beta c\gamma \end{bmatrix} \end{matrix} & \begin{matrix} c\beta c\gamma \\ +\varepsilon\begin{bmatrix} -B s\beta c\gamma \\ - G c\beta s\gamma \end{bmatrix} \end{matrix} \end{bmatrix} \tag{100} $$

Note that the expression for the transformation through Euler dual angles about the *x*-, *y*- and *z*-axes given by eq.(100) is identical to the matrix in eq.(86) representing the same three screw motions taken about fixed axes in the opposite order.

The three Euler dual angles represented in a given 3×3 dual-number matrix can be determined in the same manner as the dual roll/pitch/yaw angles for screw motions with respect to a fixed frame.

Relative Transformation

Successive coordinate transformations are concatenated into a single matrix by pre-multiplication (left multiply) if the transformations are made with respect to the axes of a fixed coordinate frame.

Successive coordinate transformations are concatenated into a single matrix by post-multiplication (right multiply) if the transformations are made with respect to the axes of a moving coordinate frame.

Example

Consider a base (fixed) frame {XYZ} and a moving frame {C} specified with respect to frame {XYZ} as

$$\{C\} = \begin{bmatrix} 0 & 1 & \varepsilon 5 \\ -1 & 0 & -\varepsilon 2 \\ -\varepsilon 2 & -\varepsilon 5 & 1 \end{bmatrix}$$

Transform frame {C} through a dual angle 90°+ε2 about the x-axis of the fixed frame {XYZ}.

$$\hat{X}(90°+\varepsilon 2) = \begin{bmatrix} 1 & 0 & 0 \\ 0 & -\varepsilon 2 & -1 \\ 0 & 1 & -\varepsilon 2 \end{bmatrix}$$

followed by a transformation through a dual angle 90°+ε3 about the z-axis (same fixed frame).

$$\hat{Z}(90°+\varepsilon 3) = \begin{bmatrix} -\varepsilon 3 & -1 & 0 \\ 1 & -\varepsilon 3 & 0 \\ 0 & 0 & 1 \end{bmatrix}$$

The final situation of the frame {B} is

$$\{B\} = \hat{Z}\hat{X}\{C\}$$

$$= \begin{bmatrix} -\varepsilon 3 & -1 & 0 \\ 1 & -\varepsilon 3 & 0 \\ 0 & 0 & 1 \end{bmatrix} \begin{bmatrix} 1 & 0 & 0 \\ 0 & -\varepsilon 2 & -1 \\ 0 & 1 & -\varepsilon 2 \end{bmatrix} \begin{bmatrix} 0 & 1 & \varepsilon 5 \\ -1 & 0 & -\varepsilon 2 \\ -\varepsilon 2 & -\varepsilon 5 & 1 \end{bmatrix}$$

$$\{B\} = \begin{bmatrix} -\varepsilon 4 & -\varepsilon 8 & 1 \\ 0 & 1 & \varepsilon 8 \\ -1 & 0 & -\varepsilon 4 \end{bmatrix}$$

Now, consider the transformation of frame {C} through dual angle 90°+ε2 about the x-axis of frame {C} and then followed by a transformation through dual angle 90°+ε3 about the z-axis of the moved frame.

$$\{B\} = \{C\}\hat{X}\hat{Z}$$

$$= \begin{bmatrix} 0 & 1 & \varepsilon 5 \\ -1 & 0 & -\varepsilon 2 \\ -\varepsilon 2 & -\varepsilon 5 & 1 \end{bmatrix} \begin{bmatrix} 1 & 0 & 0 \\ 0 & -\varepsilon 2 & -1 \\ 0 & 1 & -\varepsilon 2 \end{bmatrix} \begin{bmatrix} -\varepsilon 3 & -1 & 0 \\ 1 & -\varepsilon 3 & 0 \\ 0 & 0 & 1 \end{bmatrix}$$

$$\{B\} = \begin{bmatrix} -\varepsilon 3 & 0 & -1 \\ \varepsilon & 1 & 0 \\ 1 & -\varepsilon & \varepsilon 3 \end{bmatrix}$$

Compare these results. The same transformations taken in the same order about fixed and moving frames yield different results.

ZYZ Euler Angles

A popular form of Euler angles is the ZYZ sequence. In this scheme we start with a moving system aligned with a stationary frame {A}. First we make a screw motion with respect to the z-axis through a dual angle $\hat{\alpha}$. Next we make a screw motion with respect to the y-axis of the moving coordinate system through a dual angle $\hat{\beta}$. The we make a screw motion through a dual angle $\hat{\gamma}$ with respect to the z-axis of the moving system.

$$ {}^{A}_{B}\hat{D} = \hat{Z}(\hat{\alpha})\hat{Y}(\hat{\beta})\hat{Z}(\hat{\gamma}) \tag{101}$$

$$= \begin{bmatrix} c\hat{\alpha} & -s\hat{\alpha} & 0 \\ s\hat{\alpha} & c\hat{\alpha} & 0 \\ 0 & 0 & 1 \end{bmatrix} \begin{bmatrix} c\hat{\beta} & 0 & s\hat{\beta} \\ 0 & 1 & 0 \\ -s\hat{\beta} & 0 & c\hat{\beta} \end{bmatrix} \begin{bmatrix} c\hat{\gamma} & -s\hat{\gamma} & 0 \\ s\hat{\gamma} & c\hat{\gamma} & 0 \\ 0 & 0 & 1 \end{bmatrix} \tag{102}$$

$$= \begin{bmatrix} c\hat{\alpha}c\hat{\beta}c\hat{\gamma} - s\hat{\alpha}s\hat{\gamma} & -c\hat{\alpha}c\hat{\beta}s\hat{\gamma} - s\hat{\alpha}c\hat{\gamma} & c\hat{\alpha}s\hat{\beta} \\ s\hat{\alpha}c\hat{\beta}c\hat{\gamma} - c\hat{\alpha}s\hat{\gamma} & -s\hat{\alpha}c\hat{\beta}s\hat{\gamma} + c\hat{\alpha}c\hat{\gamma} & s\hat{\alpha}s\hat{\beta} \\ -s\hat{\beta}c\hat{\gamma} & s\hat{\beta}s\hat{\gamma} & c\hat{\beta} \end{bmatrix} \tag{103}$$

$$
{}^{A}_{B}\hat{\mathrm{D}} = \begin{bmatrix}
\begin{array}{c} -s\alpha s\gamma + c\alpha c\beta c\gamma \\ -\varepsilon\begin{bmatrix} A(c\alpha s\gamma - s\alpha c\beta c\gamma) \\ +Bc\alpha s\beta c\gamma \\ +G(s\alpha c\gamma + c\alpha c\beta s\gamma) \end{bmatrix} \end{array} &
\begin{array}{c} -s\alpha c\gamma - c\alpha c\beta s\gamma \\ +\varepsilon\begin{bmatrix} -A(c\alpha c\gamma - s\alpha c\beta s\gamma) \\ +Bc\alpha s\beta s\gamma \\ +G(s\alpha s\gamma - c\alpha c\beta c\gamma) \end{bmatrix} \end{array} &
\begin{array}{c} c\alpha s\beta \\ +\varepsilon\begin{bmatrix} -As\alpha s\beta \\ +Bc\alpha c\beta \end{bmatrix} \end{array} \\
\begin{array}{c} c\alpha s\gamma + s\alpha c\beta c\gamma \\ +\varepsilon\begin{bmatrix} A(s\alpha s\gamma + c\alpha c\beta c\gamma) \\ -Bs\alpha s\beta c\gamma \\ -G(c\alpha c\gamma - s\alpha c\beta s\gamma) \end{bmatrix} \end{array} &
\begin{array}{c} c\alpha c\gamma - s\alpha c\beta s\gamma \\ +\varepsilon\begin{bmatrix} -A(s\alpha c\gamma + c\alpha c\beta s\gamma) \\ +Bs\alpha s\beta s\gamma \\ -G(c\alpha c\gamma + s\alpha c\beta s\gamma) \end{bmatrix} \end{array} &
\begin{array}{c} s\alpha s\beta \\ +\varepsilon\begin{bmatrix} Ac\alpha s\beta \\ +Bs\alpha c\beta \end{bmatrix} \end{array} \\
\begin{array}{c} -s\beta c\gamma \\ +\varepsilon\begin{bmatrix} -Bc\beta c\gamma \\ +Gs\beta s\gamma \end{bmatrix} \end{array} &
\begin{array}{c} s\beta s\gamma \\ +\varepsilon\begin{bmatrix} Bc\beta s\gamma \\ +Gs\beta c\gamma \end{bmatrix} \end{array} &
\begin{array}{c} c\beta \\ -\varepsilon Bs \end{array}
\end{bmatrix} \tag{104}
$$

From eq.(104) it is observed that if $\sin\beta \neq 0$ (the solution degenerates when $\beta = 0, \pi$)

$$\alpha = \tan^{-1}\frac{d_{23p}}{d_{13p}} \tag{105}$$

$$\beta = \tan^{-1}\frac{\sqrt{d_{31p}^2 + d_{32p}^2}}{d_{33p}} \tag{106}$$

$$\gamma = \tan^{-1}\frac{d_{32p}}{\left(-d_{31p}\right)} \tag{107}$$

and

$$B = -\frac{d_{33d}}{s\beta} \tag{108}$$

$$A = \frac{d_{32d} - Bs\alpha c\beta}{c\alpha s\beta} \tag{109}$$

$$G = \frac{d_{32d} - Bc\beta s\gamma}{s\beta c\gamma} \tag{110}$$

Screw Motion With Respect to a General Unit Vector

The coordinate-transformation matrices which have been developed deal with screw motions only with respect to the axes of coordinate frames. However, we may wish to consider screw motions with respect to a general axis, some line in space not coincident with the coordinate axes.

Suppose that we wish to transform from a coordinate frame {A} into coincidence with a coordinate frame {B} by taking a screw motion through dual angle $\hat{\theta} = \theta + \varepsilon s$ with respect to some axis represented by the unit line vector $\hat{h} = \vec{f} + \varepsilon \vec{g}$ such that

$$\{A\} = {}^{A}_{B}\hat{T}(\hat{\theta}, \hat{h})\{B\} \tag{111}$$

In order to accommodate our limitation of only being able to consider screw motions with respect to coordinate axes, the screw motion is considered about some frame {C} whose z-axis coincides with the general screw axis. Thus the coordinate frame {A} in which the screw axis $\hat{h}$ has been specified is transformed to this frame {C}. Then the screw motion of the body through dual angle $\hat{\theta}$ is taken with respect to the ${}^{C}z$-axis which corresponds to the specified screw axis $\hat{h}$. The result of this is then transformed from frame {C} back to the reference frame {A} in which the unit vectors of frame {B} are to be specified. This is accomplished by the concatenation of the coordinate-transformation matrices

$$ {}^{A}_{B}\hat{T}(\hat{\theta}, \hat{h}) = {}^{A}_{B}\hat{T}\hat{Z}(\hat{\theta})\,{}^{C}_{A}\hat{T} \tag{112}$$

where matrix $\hat{Z}(\hat{\theta})$ represents a screw motion through dual angle $\hat{\theta}$ about a z-axis and matrix ${}^{A}_{C}\hat{T}$ (of which matrix ${}^{C}_{A}\hat{T}$ is the inverse) represents the transformation from coordinate frame {A} to coordinate frame {B}. Now, the three columns of transformation matrix ${}^{A}_{C}\hat{T}$ represent the dual unit vectors of frame {C} as expressed in terms of the coordinates of frame {A}. Of these, the elements of the ${}^{C}x$-axis are represented by symbols $\hat{A}, \hat{B}, \hat{C}$, the ${}^{C}y$-axis by symbols $\hat{D}, \hat{E}, \hat{F}$, while the ${}^{C}z$-axis is set in alignment with the axis of the screw motion, dual vector $\hat{h}$ so that

$$ {}^{A}_{C}\hat{T} = \begin{bmatrix} \hat{A} & \hat{D} & \hat{h}_x \\ \hat{B} & \hat{E} & \hat{h}_y \\ \hat{C} & \hat{F} & \hat{h}_z \end{bmatrix} \tag{113}$$

Consequently, eq.(112) can be rewritten as

$$ {}^{A}_{B}\hat{T} = \begin{bmatrix} \hat{A} & \hat{D} & \hat{h}_x \\ \hat{B} & \hat{E} & \hat{h}_y \\ \hat{C} & \hat{F} & \hat{h}_z \end{bmatrix} \begin{bmatrix} c\hat{\theta} & -s\hat{\theta} & 0 \\ s\hat{\theta} & c\hat{\theta} & 0 \\ 0 & 0 & 1 \end{bmatrix} \begin{bmatrix} \hat{A} & \hat{B} & \hat{C} \\ \hat{D} & \hat{E} & \hat{F} \\ \hat{h}_x & \hat{h}_y & \hat{h}_z \end{bmatrix} \tag{114}$$

The expansion of this product can be considerably simplified by making several observations.

The dot product of any row or column of a coordinate-transformation matrix with any other row or column will vanish since the vectors comprising the matrix are perpendicular and intersecting so that

$$\hat{A}\hat{D} + \hat{B}\hat{E} + \hat{C}\hat{F} = 0 \quad (115)$$

$$\hat{A}\hat{B} + \hat{D}\hat{E} + \hat{h}_x\hat{h}_y = 0 \quad (116)$$

$$\hat{A}\hat{C} + \hat{D}\hat{F} + \hat{h}_x\hat{h}_z = 0 \quad (117)$$

$$\hat{B}\hat{C} + \hat{E}\hat{F} + \hat{h}_y\hat{h}_z = 0 \quad (118)$$

The dot product of any row or column of a coordinate-transformation matrix with itself is unity since these are unit dual vectors so that

$$\hat{A}\hat{A} + \hat{B}\hat{B} + \hat{C}\hat{C} = 1 \quad (119)$$

$$\hat{D}\hat{D} + \hat{E}\hat{E} + \hat{F}\hat{F} = 1 \quad (120)$$

$$\hat{A}\hat{A} + \hat{D}\hat{D} + \hat{h}_x\hat{h}_x = 1 \quad (121)$$

$$\hat{B}\hat{B} + \hat{E}\hat{E} + \hat{h}_y\hat{h}_y = 1 \quad (122)$$

$$\hat{C}\hat{C} + \hat{F}\hat{F} + \hat{h}_z\hat{h}_z = 1 \quad (123)$$

The cross product of any two columns of a coordinate-transformation matrix is the third column since the columns represent a set of perpendicular unit vectors intersecting at the origin of the frame they specify so that

$$\hat{B}\hat{F} - \hat{C}\hat{E} = \hat{h}_x \quad (124)$$

$$\hat{C}\hat{D} - \hat{A}\hat{F} = \hat{h}_y \quad (125)$$

$$\hat{A}\hat{E} - \hat{B}\hat{D} = \hat{h}_z \quad (126)$$

Substituting eqs.(115-126) into the expansion of eq.(114) yields the transformation matrix representing a screw motion through dual angle $\hat{\theta}$ with respect to an axis specified by unit vector $\hat{\mathrm{h}}$

$$
{}^{A}_{B}\hat{\mathrm{T}}\left(\hat{\theta},\hat{h}\right) = \begin{bmatrix} \hat{t}_{11} & \hat{t}_{12} & \hat{t}_{13} \\ \hat{t}_{21} & \hat{t}_{22} & \hat{t}_{23} \\ \hat{t}_{31} & \hat{t}_{32} & \hat{t}_{33} \end{bmatrix}
$$

$$
= \begin{bmatrix} \hat{h}_x^2\,\mathrm{v}\hat{\theta} + \mathrm{c}\hat{\theta} & \hat{h}_x\hat{h}_y\,\mathrm{v}\hat{\theta} - \hat{h}_z\,\mathrm{s}\hat{\theta} & \hat{h}_x\hat{h}_z\,\mathrm{v}\hat{\theta} + \hat{h}_y\,\mathrm{s}\hat{\theta} \\ \hat{h}_x\hat{h}_y\,\mathrm{v}\hat{\theta} + \hat{h}_z\,\mathrm{s}\hat{\theta} & \hat{h}_y^2\,\mathrm{v}\hat{\theta} + \mathrm{c}\hat{\theta} & \hat{h}_y\hat{h}_z\,\mathrm{v}\hat{\theta} - \hat{h}_x\,\mathrm{s}\hat{\theta} \\ \hat{h}_x\hat{h}_z\,\mathrm{v}\hat{\theta} - \hat{h}_y\,\mathrm{s}\hat{\theta} & \hat{h}_y\hat{h}_z\,\mathrm{v}\hat{\theta} + \hat{h}_x\,\mathrm{s}\hat{\theta} & \hat{h}_z^2\,\mathrm{v}\hat{\theta} + \mathrm{c}\hat{\theta} \end{bmatrix} \tag{127}
$$

where $\mathrm{v}\hat{\theta}$ denotes $\mathrm{versine}\,\hat{\theta} = 1 - \mathrm{c}\hat{\theta}$.

Determination of the Dual Angle and Axis of a Given Screw Motion

The Case of a General Motion

Consider the situation where it is desired to determine the dual angle $\hat{\theta} = \theta + \varepsilon s$ and $\hat{\mathrm{h}} = \vec{\mathrm{f}} + \varepsilon\vec{\mathrm{g}}$ of a general screw motion represented by a given transformation ${}^{A}_{B}\hat{\mathrm{T}}$ with elements $\hat{t}_{ij}$.

We equate the sums of the diagonal elements of eq.(127)

$$
\hat{t}_{11} + \hat{t}_{22} + \hat{t}_{33} = \hat{h}_x^2\,\mathrm{v}\hat{\theta} + \mathrm{c}\hat{\theta} + \hat{h}_y^2\,\mathrm{v}\hat{\theta} + \mathrm{c}\hat{\theta} + \hat{h}_z^2\,\mathrm{v}\hat{\theta} + \mathrm{c}\hat{\theta} = \left(\hat{h}_x^2 + \hat{h}_y^2 + \hat{h}_z^2\right)\mathrm{v}\hat{\theta} + 3\mathrm{c}\hat{\theta} \tag{128}
$$

Let

$$
\hat{t}_{11} = u_{11} + \varepsilon v_{11} \tag{129}
$$

$$
\hat{t}_{22} = u_{22} + \varepsilon v_{22} \tag{130}
$$

$$
\hat{t}_{33} = u_{33} + \varepsilon v_{33} \tag{131}
$$

and

$$
\hat{h}_x = f_x + \varepsilon g_x \tag{132}
$$

$$
\hat{h}_y = f_y + \varepsilon g_y \tag{133}
$$

$$
\hat{h}_z = f_z + \varepsilon g_z \tag{134}
$$

and

$$
\mathrm{v}\hat{\theta} = 1 - \mathrm{c}\hat{\theta} = 1 - \mathrm{c}\theta + \varepsilon s\mathrm{s}\theta \tag{135}
$$

so that eq.(128) can be written

$$u_{11} + u_{22} + u_{33} + \varepsilon(v_{11} + v_{22} + v_{33}) \\ = \\ \left[\left(f_x^2 + f_y^2 + f_z^2\right) + \varepsilon 2\left(f_x g_x + f_y g_y + f_z g_z\right)\right]\left[(1 - c\theta) + \varepsilon ss\theta\right] + 3c\theta - \varepsilon 3ss\theta \tag{136}$$

Because vector $\vec{f}$ is of unit length

$$f_x^2 + f_y^2 + f_z^2 = 1 \tag{137}$$

and so

$$u_{11} + u_{22} + u_{33} + \varepsilon(v_{11} + v_{22} + v_{33}) \\ = \\ 1 + 2c\theta + \varepsilon 2\left[\left(f_x g_x + f_y g_y + f_z g_z\right)(1 - c\theta) - ss\theta\right] \tag{138}$$

Thus the primary component of eq.(138) yields

$$u_{11} + u_{22} + u_{33} = 1 + 2c\theta \tag{139}$$

and thus

$$c\theta = \frac{u_{11} + u_{22} + u_{33} - 1}{2} \tag{140}$$

Now, considering the differences between off-diagonal terms of eq.(127)

$$\hat{t}_{32} - \hat{t}_{23} = \left(\hat{h}_y \hat{h}_z \, v\hat{\theta} + \hat{h}_x \, s\hat{\theta}\right) - \left(\hat{h}_y \hat{h}_z \, v\hat{\theta} - \hat{h}_x \, s\hat{\theta}\right) = 2\hat{h}_x \, s\hat{\theta} \tag{141}$$

The components of this equation are

$$u_{32} - u_{23} = 2f_x \, s\theta \tag{142}$$

and

$$v_{32} - v_{23} = 2\left(f_x s c\theta + g_x \, s\theta\right) \tag{143}$$

Similarly

$$u_{13} - u_{31} = 2f_y \, s\theta \tag{144}$$

$$v_{13} - v_{31} = 2\left(f_y s c\theta + g_y \, s\theta\right) \tag{145}$$

$$u_{21} - u_{12} = 2f_z \, s\theta \tag{146}$$

$$v_{12} - v_{21} = 2\left(f_z s c\theta + g_z \, s\theta\right) \tag{147}$$

An expression for the sine of the rotation angle θ is obtained by adding the squares of the primary components

$$(u_{32}-u_{23})^2+(u_{13}-u_{31})^2+(u_{21}-u_{12})^2=4\left(f_x^2+f_y^2+f_z^2\right)s^2\theta \quad (148)$$

and using eq.(137),

$$(u_{32}-u_{23})^2+(u_{13}-u_{31})^2+(u_{21}-u_{12})^2=4\,s^2\theta \quad (149)$$

so that

$$s\theta=\pm\frac{\sqrt{(u_{32}-u_{23})^2+(u_{13}-u_{31})^2+(u_{21}-u_{12})^2}}{2} \quad (150)$$

Eq.(150) evaluated for the sine of the rotation angle θ and eq.(140) evaluated for the cosine of the rotation angle θ are considered together to unambiguously (as to quadrant) determine the rotation angle θ. Usually the plus sign in eq.(150) is chosen as a matter of convenience. Choosing the negative sign would only result in a corresponding change in the sense of the unit resultant vector component of the screw axis of the motion.

Knowing the rotation angle θ, the primary component of the differences in the off-diagonal terms in eq.(127) can be used to determine the components of the unit resultant vector $\vec{f}$.

$$f_x=\frac{u_{32}-u_{23}}{2s\theta} \quad (151)$$

$$f_y=\frac{u_{13}-u_{31}}{2s\theta} \quad (152)$$

$$f_z=\frac{u_{21}-u_{12}}{2s\theta} \quad (153)$$

Now, the dot product of a resultant vector and a moment vector comprising a dual vector will vanish (which can be proven by expanding $\vec{f}\cdot\vec{g}=\vec{f}\cdot\left(\vec{r}\times\vec{f}\right)$ where $\vec{r}$ is the vector from the origin of the coordinate frame to any point on the line of the resultant vector).

$$f_x g_x+f_y g_y+f_z g_z=0 \quad (154)$$

The dual component of eq.(138) is

$$v_{11}+v_{22}+v_{33}=2\left[\left(f_x g_x+f_y g_y+f_z g_z\right)(1-c\theta)-ss\theta\right] \quad (155)$$

So, substituting eq.(154) into eq.(155),

$$v_{11}+v_{22}+v_{33}=-2ss\theta \quad (156)$$

and so the translation along the screw axis is

$$s = -\frac{v_{11} + v_{22} + v_{33}}{2\,\mathrm{s}\theta} \tag{157}$$

Eqs.(143,145,147) can now be solved for the components of the moment vector $\vec{g}$

$$g_x = \frac{v_{32} - v_{23} - 2f_x s\mathrm{c}\theta}{2\mathrm{s}\theta} \tag{158}$$

$$g_y = \frac{v_{13} - v_{31} - 2f_y s\mathrm{c}\theta}{2\mathrm{s}\theta} \tag{159}$$

$$g_z = \frac{v_{21} - v_{12} - 2f_z s\mathrm{c}\theta}{2\mathrm{s}\theta} \tag{160}$$

The Case of Pure Translation

When a motion is a pure translation, when the rotation angle $\theta = 0°$, eqs.(151,152,153) cannot be used to evaluate the direction of the motion. This case is readily detected since for when rotation angle $\theta = 0°$, eq.(139) infers that

$$u_{11} + u_{22} + u_{33} = 3 \tag{161}$$

It then remains to determine the translation distance s and the direction of the motion which is given by the resultant vector $\vec{f}$.

When rotation angle $\theta = 0°$, eq.(127) becomes

$$\begin{bmatrix} \hat{t}_{11} & \hat{t}_{12} & \hat{t}_{13} \\ \hat{t}_{21} & \hat{t}_{22} & \hat{t}_{23} \\ \hat{t}_{31} & \hat{t}_{32} & \hat{t}_{33} \end{bmatrix} = \begin{bmatrix} 1 & -\varepsilon s f_z & \varepsilon s f_y \\ \varepsilon s f_z & 1 & -\varepsilon s f_x \\ -\varepsilon s f_y & \varepsilon s f_x & 1 \end{bmatrix} \tag{162}$$

of which the dual component is

$$\begin{bmatrix} v_{11} & v_{12} & v_{13} \\ v_{21} & v_{22} & v_{23} \\ v_{31} & v_{32} & v_{33} \end{bmatrix} = \begin{bmatrix} 0 & -sf_z & sf_y \\ sf_z & 0 & -sf_x \\ -sf_y & sf_x & 0 \end{bmatrix} \tag{163}$$

Adding the squares of the 3-2, 1-3 and 2-1 elements and observing that resultant vector $\vec{f}$ is of unit length

$$v_{32}^2 + v_{13}^2 + v_{21}^2 = s^2 f_x^2 + s^2 f_y^2 + s^2 f_z^2 = s^2 \tag{164}$$

Thus the translation is

$$s = \pm\sqrt{v_{32}^2 + v_{13}^2 + v_{21}^2} \tag{165}$$

The choice of one sign or the other in the translation s will result in a corresponding change in the computed sense of the resultant vector $\vec{\mathrm{f}}$ so to maintain the same motion.

When the amount of the translation s is known, the components of the resultant vector $\vec{\mathrm{f}}$ giving the direction of the translation can be determined from the 3-2, 1-3 and 2-1 elements of eq.(163)

$$f_x = \frac{v_{32}}{s} \tag{166}$$

$$f_y = \frac{v_{13}}{s} \tag{167}$$

$$f_z = \frac{v_{21}}{s} \tag{168}$$

Now, the axis of a pure translation is a unique direction by not a unique line as would be the axis of a general screw motion or a pure translation. Consequently for a pure translation the moment vector $\vec{\mathrm{g}}$ is meaningless, and need not be calculated.

The Case of a Half-Turn Screw Motion

Another situation in which eqs.(151,152,153) cannot be used is when the rotation of a screw motion is a half turn, when rotation angle θ=180°. For the case of a half turn, eq.(139) infers that

$$u_{11} + u_{22} + u_{33} = -1 \tag{169}$$

This condition can be used to readily detect the situation of a half turn. It then remains to determine the resultant vector $\vec{\mathrm{f}}$ indicating the direction and the moment vector $\vec{\mathrm{g}}$ inferring the location of the screw axis and the amount of translation s.

When the rotation of a screw motion is a half turn, eq.(127) becomes

$$\begin{bmatrix} \hat{t}_{11} & \hat{t}_{12} & \hat{t}_{13} \\ \hat{t}_{21} & \hat{t}_{22} & \hat{t}_{23} \\ \hat{t}_{31} & \hat{t}_{32} & \hat{t}_{33} \end{bmatrix} = \begin{bmatrix} 2f_x^2 - 1 + \varepsilon 4 f_x g_x & 2f_x f_y + \varepsilon\left(2f_x g_y + 2f_y g_x + s f_z\right) & 2f_x f_z + \varepsilon\left(2f_x g_z + 2f_z g_x - s f_y\right) \\ 2f_x f_y + \varepsilon\left(2f_x g_y + 2f_y g_x - s f_z\right) & 2f_y^2 - 1 + \varepsilon 4 f_y g_y & 2f_y f_z + \varepsilon\left(2f_y g_z + 2f_z g_y + s f_x\right) \\ 2f_x f_z + \varepsilon\left(2f_x g_z + 2f_z g_x + s f_y\right) & 2f_z f_y + \varepsilon\left(2f_y g_z + 2f_z g_y - s f_x\right) & 2f_z^2 - 1 + \varepsilon 4 f_z g_z \end{bmatrix} \tag{170}$$

The primary component of this equation is

$$\begin{bmatrix} u_{11} & u_{12} & u_{13} \\ u_{21} & u_{22} & u_{23} \\ u_{31} & u_{32} & u_{33} \end{bmatrix} = \begin{bmatrix} 2f_x^2 - 1 & 2f_x f_y & 2f_x f_z \\ 2f_x f_y & 2f_y^2 - 1 & 2f_y f_z \\ 2f_x f_z & 2f_z f_y & 2f_z^2 - 1 \end{bmatrix} \tag{171}$$

The 1-1, 1-2 and 1-3 elements of eq.(171) can be used to determine the components of the resultant vector $\vec{f}$

$$f_x = \pm\sqrt{\frac{u_{11} + 1}{2}} \tag{172}$$

$$f_y = \frac{u_{12}}{2f_x} \tag{173}$$

$$f_z = \frac{u_{13}}{2f_x} \tag{174}$$

The dual component of eq.(170) is

$$\begin{bmatrix} v_{11} & v_{12} & v_{13} \\ v_{21} & v_{22} & v_{23} \\ v_{31} & v_{32} & v_{33} \end{bmatrix} = \begin{bmatrix} 4f_x g_x & 2\left(f_x g_y + f_y g_x\right) + s f_z & 2\left(f_x g_z + f_z g_x\right) - s f_y \\ 2\left(f_x g_y + f_y g_x\right) - s f_z & 4 f_y g_y & 2\left(f_y g_z + f_z g_y\right) + s f_z \\ 2\left(f_x g_z + f_z g_x\right) + s f_y & 2\left(f_y g_z + f_z g_y\right) - s f_x & 4 f_z g_z \end{bmatrix} \tag{175}$$

Using the results obtained for the components of the resultant vector $\vec{f}$, the diagonal components of eq.(175) can be used to determine the components of the moment vector $\vec{g}$

$$g_x = \frac{v_{11}}{4f_x} \tag{176}$$

$$g_y = \frac{v_{22}}{4f_y} \tag{177}$$

$$g_z = \frac{v_{33}}{4f_z} \tag{178}$$

An off-diagonal element of eq.(175) can be used to determine the translation s.

$$s = \frac{v_{12} - 2\left(f_x g_y + f_y g_x\right)}{f_z} \tag{179}$$

or

$$s = \frac{v_{23} - 2\left(f_y g_z + f_z g_y\right)}{f_x} \tag{180}$$

or

$$s = \frac{v_{31} - 2\left(f_x g_z + f_z g_x\right)}{f_y} \tag{181}$$

The choice of eqs.(179,180,181) is determined by the largest of the resultant vector components f_x, f_y or f_z so that the division is performed with the most accuracy.

Example

It is desired to determine the dual angle and screw axis of the motion represented by the 3×3 dual-number coordinate-transformation matrix

$$\begin{bmatrix} -0.6077 & -0.2995 & 0.7355 \\ 0.7942 & -0.2367 & 0.5597 \\ 0.0065 & 0.9243 & 0.3817 \end{bmatrix} + \varepsilon \begin{bmatrix} -4.9576 & 5.6075 & -1.8127 \\ -3.8538 & -7.3523 & 2.3589 \\ 7.4113 & -0.0658 & 0.0338 \end{bmatrix}$$

Now, the sum of the diagonal elements of the primary component is −0.4627 which infers that this is the case of a general screw motion since that value lies between the extreme of −1 representing a half turn and the other bound of 3 which would infer a pure translation. The rotation angle θ is then computed using elements of the primary component of the given transformation matrix in eqs.(140) and (150), arriving at the result θ = 137°. Then eqs.(151,152,153) are employed with that result and elements of the primary component of the given matrix to calculate the resultant unit vector of the screw axis which is found to be

$\vec{f} = 0.2673\vec{i} + 0.5345\vec{j} + 0.8018\vec{k}$. The amount of translation along the screw axis is found to be $s = 9$ by substituting the amount of rotation and certain elements of the dual component of the transformation matrix into eq.(157). Then the moment vector of the screw axis, $\vec{g} = 0.8018\vec{i} - 1.6036\vec{j} + 0.018\vec{k}$ is obtained by substituting the amount of rotation and elements of the dual component of the transformation matrix into eqs.(158,159,160).

Exercises - Coordinate Transformation

1. What is the result when a vector $5\vec{i} + 4\vec{j} - 7\vec{k}$ passing through point $\{2,9,8\}^T$ is subjected to a screw motion of 45° and 7 length units with respect to the z-axis?

2. What is the coordinate-transformation matrix which represents a rotation of 30° about the x-axis followed by a translation of 5 length units along the y-axis of the same coordinate frame? What is the result if the order of the motions is reversed?

3. What is the transformation matrix for a screw motion of 45° and 15 length units about the x-axis, followed by a screw motion of 60° and 3 length units about the y-axis of the same coordinate frame? Calculate the result if the motions are made about moving coordinate frames.

4. What is the transformation matrix representing successive motions with respect to fixed axes of dual roll 120° and 8 length units, dual pitch 33° and 7 length units and dual yaw 60° and 4 length units?

5. Obtain the coordinate-transformation matrix for displacement through ZYX Euler dual angles of $\hat{\alpha} = 45° + \varepsilon 23$, $\hat{\beta} = 180° - \varepsilon 3$, $\hat{\gamma} = 30° + \varepsilon 12$.

6. What are the ZYX Euler dual angles corresponding to the following coordinate-transformation matrix?

$$\begin{bmatrix} 0.433 + \varepsilon 12 & 0.5 - \varepsilon 7.956 & -0.75 - \varepsilon 12.232 \\ 0.25 - \varepsilon 1.732 & -0.866 - \varepsilon 4.593 & -0.433 + \varepsilon 8.187 \\ -0.866 - \varepsilon 6.5 & -\varepsilon 10.5 & -0.5 + \varepsilon 11.258 \end{bmatrix}$$

7. What is the dual angle and the axis of the screw motion specified by the following coordinate-transformation matrix?

$$\begin{bmatrix} -0.5535 & 0.8063 & 0.2085 \\ 0.2489 & 0.3991 & -0.8825 \\ -0.7948 & -0.4366 & -0.4216 \end{bmatrix} + \varepsilon \begin{bmatrix} -20.1002 & -12.1609 & -6.3317 \\ -8.0367 & 0.3259 & -2.1194 \\ 11.4818 & -22.1629 & 1.3044 \end{bmatrix}$$

Chapter 3 - Matrix Modeling of Joints and Links

Definitions

A mechanism is comprised of members called "links" whose connections are called "joints."

On each link in a mechanism we fix a coordinate frame. Using the mathematics of coordinate transformation, we can determine the relationship of each frame to the other and thus determine the position and orientation of each link as the mechanism articulates.

A link is considered as a rigid body defining the relationship between two neighboring joint axes. The two joint axes of a link are each a line in space. Two parameters define their relative locations:

a = distance between the joint axes of a link (link length)

α = angle between the joint axes of a link (link twist)

Thus the dimensions of a link can be defined by a dual angle

$$\hat{\alpha} = \alpha + \varepsilon a \tag{182}$$

Neighboring links have a joint between them. The joint parameters are:

θ = joint rotation (angular motion)

s = joint offset (sliding, translation or prismatic motion)

Thus, the displacement at a joint may be specified by a dual angle

$$\hat{\theta} = \theta + \varepsilon s \tag{183}$$

The values α, a, θ, s are often referred to as "Denavit-Hartenberg parameters."

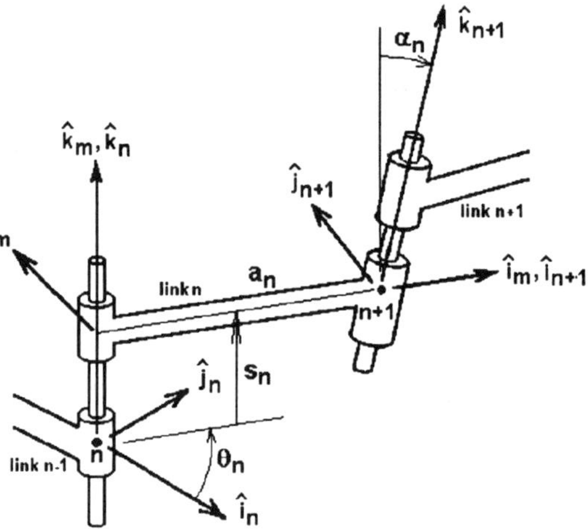

Figure 11 Link n with joints n and n-1. Frame {n} is fixed on the previous link n-1. Angle θ_n represents the rotation of link n relative to link n-1. Translation s_n is the distance from the point n, the origin of frame {n} to the line segment whose length a_n is the shortest distance between joint axes $\hat{k}_n$ and $\hat{k}_{n+1}$. That line segment of shortest distance between joint axes intersects axis $\hat{k}_{n+1}$ at point n+1, the origin of frame {n+1} fixed on the distal end of link n. The projected angle between axes $\hat{k}_n$ and $\hat{k}_{n+1}$ represents the twist α_n of link n.

Frame Location

Mechanism analysis is facilitated by fixing a coordinate frame on each link in a specific manner. In this book, a convention is adopted, as shown in Figure 11, where a coordinate frame {n+1} is fixed on the distal end of a link n joining joints n and n+1 such that

$\hat{k}_{n+1}$ axis coincident with axis of joint n+1,

$\hat{i}_{n+1}$ axis coincident with shortest distance between axes $\hat{k}_n$ and $\hat{k}_{n+1}$,

$\hat{j}_{n+1}$ axis is perpendicular to both axes $\hat{i}_{n+1}$ and $\hat{k}_{n+1}$,

The origin of frame {n+1} is located at the intersection of axes $\hat{k}_{n+1}$ and $\hat{i}_{n+1}$.

Joint-Link Transformations

The coordinate-transformation matrix modeling each link of a mechanism will be considered the product of a matrix specified by the displacements at the proximal joint of the link and a matrix specified by the dimensions of the link.

Each of these matrices will be a function of a single dual-number parameter only, the scalar components of which are readily obtained by inspection.

Let us trace the path through the link as shown in Figure 11, from frame {n} fixed on link n-1 at joint n to frame {n+1} fixed on link n at joint n+1.

First, transform through dual angle $\hat{\theta}_n = \theta_n + \varepsilon s_n$ about axis $\hat{k}_n$ from frame {n} to frame {m}.

$$ {}^{n}_{m}\hat{Z}(\hat{\theta}_n) = \begin{bmatrix} c\hat{\theta}_n & -s\hat{\theta}_n & 0 \\ s\hat{\theta}_n & c\hat{\theta}_n & 0 \\ 0 & 0 & 1 \end{bmatrix} \tag{184} $$

Secondly, transform through dual angle $\hat{\alpha} = \alpha + \varepsilon a$ about axis $\hat{i}_m$ from frame {m} to frame {n+1}.

$$ {}^{m}_{n+1}\hat{X}(\hat{\alpha}_n) = \begin{bmatrix} 1 & 0 & 0 \\ 0 & c\hat{\alpha}_n & -s\hat{\alpha}_n \\ 0 & s\hat{\alpha}_n & c\hat{\alpha}_n \end{bmatrix} \tag{185} $$

We concatenate to form the total expression modeling the link.

$$ {}^{n}_{n+1}\hat{M}(\hat{\theta}_n, \hat{\alpha}_n) = {}^{n}_{n+1}\hat{Z}(\hat{\theta}_n)\,{}^{n}_{n+1}\hat{X}(\hat{\alpha}_n) \tag{186} $$

$$ = \begin{bmatrix} c\hat{\theta}_n & -s\hat{\theta}_n & 0 \\ s\hat{\theta}_n & c\hat{\theta}_n & 0 \\ 0 & 0 & 1 \end{bmatrix} \begin{bmatrix} 1 & 0 & 0 \\ 0 & c\hat{\alpha}_n & -s\hat{\alpha}_n \\ 0 & s\hat{\alpha}_n & c\hat{\alpha}_n \end{bmatrix} \tag{187} $$

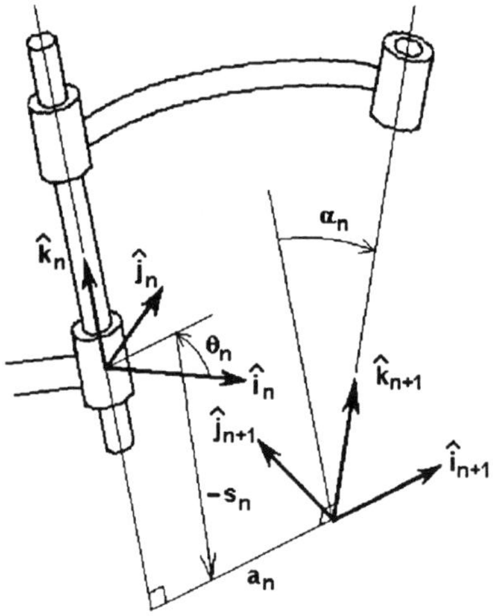

Figure 12 While a link may be arranged in an arch, the definitions of its dimensions are based on the same definitions as for the straight link.

$$= \begin{bmatrix} c\hat{\theta}_n & -c\hat{\alpha}_n s\hat{\theta}_n & s\hat{\alpha}_n s\hat{\theta}_n \\ s\hat{\theta}_n & c\hat{\alpha}_n c\hat{\theta}_n & -s\hat{\alpha}_n c\hat{\theta}_n \\ 0 & s\hat{\alpha}_n & c\hat{\alpha}_n \end{bmatrix} \tag{188}$$

$$\begin{gathered} {}^{n}_{n+1}\hat{M}\left(\hat{\theta}_n, \hat{\alpha}_n\right) \\ = \\ \begin{bmatrix} c\theta_n & -c\alpha_n s\theta_n & s\alpha_n s\theta_n \\ s\theta_n & c\alpha_n c\theta_n & -s\alpha_n c\theta_n \\ 0 & s\alpha_n & c\alpha_n \end{bmatrix} \\ + \\ \varepsilon \begin{bmatrix} -s_n s\theta_n & a_n s\alpha_n s\theta_n - s_n c\alpha_n c\theta_n & a_n c\alpha_n s\theta_n + s_n s\alpha_n c\theta_n \\ s_n c\theta_n & -a_n s\alpha_n c\theta_n - s_n c\alpha_n s\theta_n & -a_n c\alpha_n c\theta_n + s_n s\alpha_n s\theta_n \\ 0 & a_n c\alpha_n & -a_n s\alpha_n \end{bmatrix} \end{gathered} \tag{189}$$

Eq.(189) is the joint-link modeling equation for any binary (two-joint) link with any of revolute, prismatic or cylindrical joints.

Observe that an arched link as shown in Figure 12, may appear different than a straight or twisted link as shown in Figure 11, but the same definitions apply. The material comprising the link does not have to be arranged about the shortest distance between the joint axes.

If the joint is revolute, then the joint offset is constant, set to the overlap between the links as shown in Figure 13. Often mechanisms are considered planar and the overlap neglected ($s_n = 0$). Sometimes a clevis construction, as shown in Figure 14, is used to obtain a true $s_n = 0$ joint.

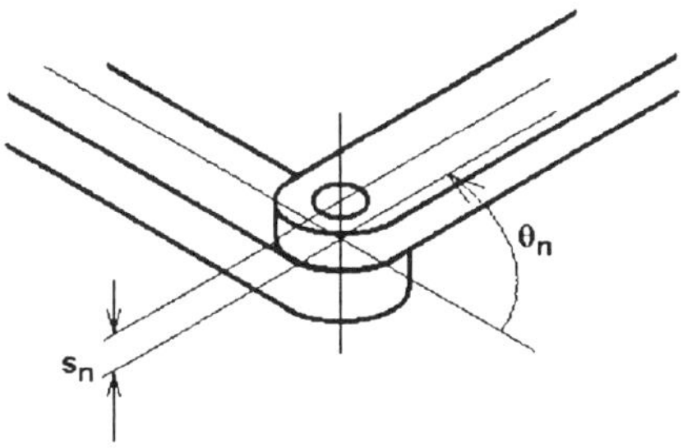

Figure 13 Overlap on account of the thickness of links results in a fixed joint translation s_n.

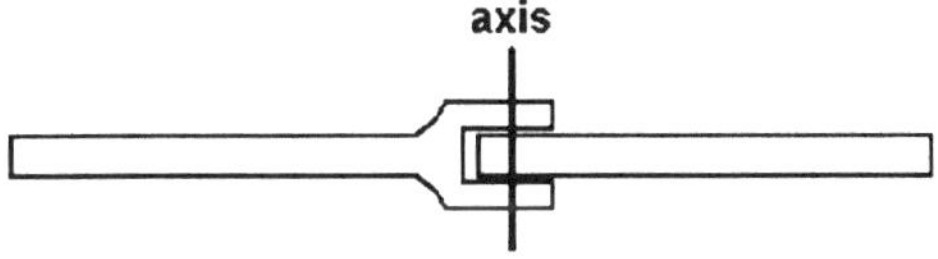

Figure 14 A clevis construction is sometimes used to obtain a revolute joint without overlap.

Links with Ball Joints

A ball joint permits motion with respect to all three of a set of mutually-perpendicular axes. However, these axes are not naturally defined as is the axis of a cylindrical joint. Therefore, we will select the axes of allowed screw motions arbitrarily, choosing ZYX Euler dual angles.

$$ {}^{n}_{n+1}\hat{L} = \hat{Z}(\hat{\theta}_n)\hat{Y}(\hat{\eta}_n)\hat{X}(\hat{\zeta}_n)\hat{X}(\hat{\alpha}_n) \tag{190}$$

$$ = \begin{bmatrix} c\hat{\theta}_n & -s\hat{\theta}_n & 0 \\ s\hat{\theta}_n & c\hat{\theta}_n & 0 \\ 0 & 0 & 1 \end{bmatrix} \begin{bmatrix} c\hat{\eta}_n & 0 & s\hat{\eta}_n \\ 0 & 1 & 0 \\ -s\hat{\eta}_n & 0 & c\hat{\eta}_n \end{bmatrix} \begin{bmatrix} 1 & 0 & 0 \\ 0 & c\hat{\zeta}_n & -s\hat{\zeta}_n \\ 0 & s\hat{\zeta}_n & c\hat{\zeta}_n \end{bmatrix} \begin{bmatrix} 1 & 0 & 0 \\ 0 & c\hat{\alpha}_n & -s\hat{\alpha}_n \\ 0 & s\hat{\alpha}_n & c\hat{\alpha}_n \end{bmatrix} \tag{191}$$

Considering the Euler dual angles $\hat{\theta}_n = \theta_n + \varepsilon S_n$, $\hat{\eta}_n = \eta_n + \varepsilon E_n$, $\hat{\zeta}_n = \zeta_n + \varepsilon Z_n$, $\hat{\alpha}_n = \alpha_n + \varepsilon A_n$, eq.(191) can be expanded as

$$
{}^{n}_{n+1}\hat{\mathrm{L}} =
\begin{bmatrix}
\begin{matrix} c\theta_n c\eta_n \\ -\varepsilon\begin{bmatrix} S_n s\theta_n c\eta_n \\ +E_n c\theta_n s\eta_n \end{bmatrix} \end{matrix} &
\begin{matrix} -s\theta_n c\zeta_n\alpha_n + c\theta_n s\eta_n s\zeta_n\alpha_n \\ +\varepsilon\begin{bmatrix} S_n(-c\theta_n c\zeta_n\alpha_n + s\theta_n s\eta_n s\zeta_n\alpha_n) \\ +E_n c\theta_n c\eta_n s\zeta_n\alpha_n \\ +(Z_n + A_n)(s\theta_n s\zeta_n\alpha_n - s\theta_n s\eta_n s\zeta_n\alpha_n) \end{bmatrix} \end{matrix} &
\begin{matrix} s\theta_n s\zeta_n\alpha_n + c\theta_n s\eta_n c\zeta_n\alpha_n \\ +\varepsilon\begin{bmatrix} -S_n c\theta_n s\zeta_n\alpha_n \\ +E_n c\theta_n c\eta_n c\zeta_n\alpha_n \\ +(Z_n + A_n)(s\theta_n c\zeta_n\alpha_n - c\theta_n s\eta_n s\zeta_n\alpha_n) \end{bmatrix} \end{matrix} \\
\begin{matrix} s\theta_n c\eta_n \\ +\varepsilon\begin{bmatrix} S_n c\theta_n c\eta_n \\ +E_n s\theta_n s\eta_n \end{bmatrix} \end{matrix} &
\begin{matrix} c\theta_n c\zeta_n\alpha_n + s\theta_n s\eta_n s\zeta_n\alpha_n \\ +\varepsilon\begin{bmatrix} S_n(-s\theta_n c\zeta_n\alpha_n + c\theta_n s\eta_n s\zeta_n\alpha_n) \\ +E_n s\theta_n c\eta_n s\zeta_n\alpha_n \\ +(Z_n + A_n)(-c\theta_n s\zeta_n\alpha_n + s\theta_n c\eta_n s\zeta_n\alpha_n) \end{bmatrix} \end{matrix} &
\begin{matrix} -c\theta_n s\zeta_n\alpha_n + s\theta_n s\eta_n c\zeta_n\alpha_n \\ +\varepsilon\begin{bmatrix} S_n(s\theta_n s\zeta_n\alpha_n + c\theta_n s\eta_n c\zeta_n\alpha_n) \\ +E_n s\theta_n c\eta_n c\zeta_n\alpha_n \\ -(Z_n + A_n)(c\theta_n c\zeta_n\alpha_n - s\theta_n s\eta_n s\zeta_n\alpha_n) \end{bmatrix} \end{matrix} \\
\begin{matrix} -s\eta_n \\ -\varepsilon E_n c\eta_n \end{matrix} &
\begin{matrix} c\eta_n s\zeta_n\alpha_n \\ +\varepsilon\begin{bmatrix} -E_n s\eta_n s\zeta_n\alpha_n \\ +(Z_n + A_n)(c\eta_n c\zeta_n\alpha_n) \end{bmatrix} \end{matrix} &
\begin{matrix} c\eta_n c\zeta_n\alpha_n \\ +\varepsilon\begin{bmatrix} -E_n s\eta_n c\zeta_n\alpha_n \\ +(Z_n + A_n)(c\eta_n s\zeta_n\alpha_n) \end{bmatrix} \end{matrix}
\end{bmatrix}
\tag{192}
$$

where $s\hat{\zeta}_n\hat{\alpha}_n = s(\hat{\zeta}_n + \hat{\alpha}_n)$ and $c\hat{\zeta}_n\hat{\alpha}_n = c(\hat{\zeta}_n + \hat{\alpha}_n)$.

Often the motion through dual angle $\hat{\zeta}_n = \zeta_n + \varepsilon Z_n$ represents a passive degree of freedom and link twist α_n is meaningless. For instance, a link connected by ball joints at both ends as shown in Figure 15 can rotate about its axis without meaningful effect on the situation of other links in the mechanism. Taking into account this simplification,

$$
{}^{n}_{n+1}\hat{\mathrm{K}} = \hat{\mathrm{Z}}(\hat{\theta}_n)\hat{\mathrm{Y}}(\hat{\eta}_n)\mathrm{X}(A_n) \tag{193}
$$

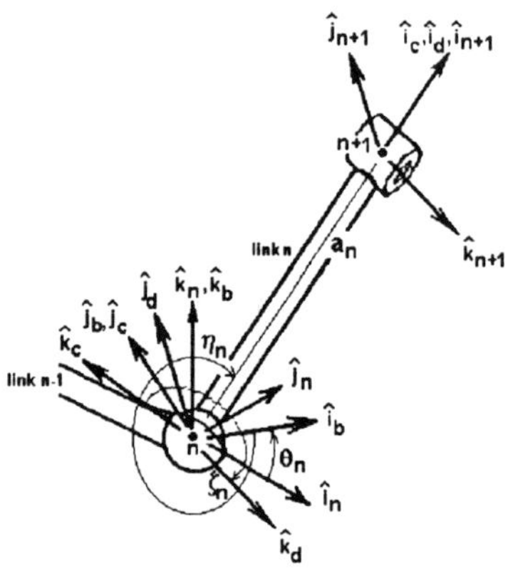

Figure 15 Link n of length a_n with ball joint which allows rotations θ_n, η_n and ζ_n.

$$= \begin{bmatrix} c\hat{\theta}_n & -s\hat{\theta}_n & 0 \\ s\hat{\theta}_n & c\hat{\theta}_n & 0 \\ 0 & 0 & 1 \end{bmatrix} \begin{bmatrix} c\hat{\eta}_n & 0 & s\hat{\eta}_n \\ 0 & 1 & 0 \\ -s\hat{\eta}_n & 0 & c\hat{\eta}_n \end{bmatrix} \begin{bmatrix} 1 & 0 & 0 \\ 0 & 1 & -\varepsilon A_n \\ 0 & \varepsilon A_n & 1 \end{bmatrix} \tag{194}$$

$$^{n}_{n+1}\hat{K} \tag{195}$$

$$= \begin{bmatrix} \begin{matrix} c\theta_n\, c\eta_n \\ -\varepsilon[S_n\, s\theta_n\, c\eta_n + E_n\, c\theta_n\, s\eta_n] \end{matrix} & \begin{matrix} -s\theta_n \\ -\varepsilon[S_n\, c\theta_n - A_n\, c\theta_n\, s\eta_n] \end{matrix} & \begin{matrix} c\theta_n\, s\eta_n \\ -\varepsilon[S_n\, s\theta_n\, s\eta_n - E_n\, c\theta_n\, c\eta_n - A_n\, s\theta_n] \end{matrix} \\ \begin{matrix} s\theta_n\, c\eta_n \\ +\varepsilon[S_n\, c\theta_n\, c\eta_n - E_n\, s\theta_n\, s\eta_n] \end{matrix} & \begin{matrix} c\theta_n \\ -\varepsilon[S_n\, s\theta_n - A_n\, s\theta_n\, s\eta_n] \end{matrix} & \begin{matrix} s\theta_n\, s\eta_n \\ +\varepsilon[S_n\, c\theta_n\, s\eta_n + E_n\, s\theta_n\, c\eta_n - A_n\, c\theta_n] \end{matrix} \\ \begin{matrix} -s\eta_n \\ -\varepsilon E_n\, c\eta_n \end{matrix} & \varepsilon A_n\, c\eta_n & \begin{matrix} c\eta_n \\ -\varepsilon E_n\, s\eta_n \end{matrix} \end{bmatrix}$$

Typically the motions allowed by a ball joint are purely rotational but screw motions have been considered to allow for generality.

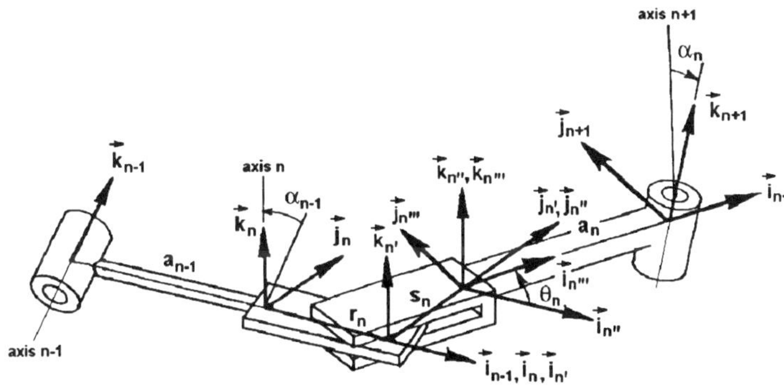

Figure 16 The plane joint.

Links with Plane Joints

A plane joint features a plane surface on one link held in contact to form a bearing with a plane surface on another link so that three degrees of freedom are allowed. These freedoms are two translations on the surfaces of the bearing planes and a rotation about an axis perpendicular to the bearing planes.

The joint on the proximal end of link n as shown in Figure 16 is modeled by considering the translation through displacement r_n along the $\vec{i}_n$ axis from frame {n} to frame {n′} followed by translation through distance s_n

along the $\vec{j}_{n'}$ to frame {n″} and then by a rotation through angle θ_n about the $\vec{k}_{n''}$ axis into alignment with frame {n‴} which is followed by a screw motion through link length a_n and twist α_n about the $\vec{i}_{n'''}$ axis to frame {n+1}. This can be represented by

$$ {}^{n}_{n+1}\hat{P} = {}^{n}_{n'}\hat{X}(\varepsilon r_n)\ {}^{n'}_{n''}\hat{Y}(\varepsilon s_n)\, {}^{n''}_{n'''}\hat{Z}(\theta_n)\ {}^{n'''}_{n+1}\hat{X}(\alpha_n + \varepsilon a_n) \tag{196}$$

$$ = \begin{bmatrix} 1 & 0 & 0 \\ 0 & 1 & -\varepsilon r_n \\ 0 & \varepsilon r_n & 1 \end{bmatrix} \begin{bmatrix} 1 & 0 & \varepsilon s_n \\ 0 & 1 & 0 \\ -\varepsilon s_n & 0 & 1 \end{bmatrix} \begin{bmatrix} c\theta_n & -s\theta_n & 0 \\ s\theta_n & c\theta_n & 0 \\ 0 & 0 & 1 \end{bmatrix} \begin{bmatrix} 1 & 0 & 0 \\ 0 & 1 & -\varepsilon a_n \\ 0 & \varepsilon a_n & 1 \end{bmatrix} \tag{197}$$

$$ {}^{n}_{n+1}P = \begin{bmatrix} c\theta_n & -s\theta_n & \varepsilon(a_n\, s\theta_n + s_n) \\ s\theta_n & c\theta_n & -\varepsilon(a_n\, c\theta_n + r_n) \\ \varepsilon(r_n\, s\theta_n - s_n\, c\theta_n) & \varepsilon(r_n\, c\theta_n + s_n\, s\theta_n + \varepsilon a_n) & 1 \end{bmatrix} \tag{198}$$

Other Conventions

The joint-link transform has been formulated in a variety of ways by researchers. It is important to understand the formulation being used and maintain consistency within any single analysis.

For instance, if the transformation through the link is formulated considering the joint on the distal end, then

$$ \hat{T}_{XZ} = \hat{X}\hat{Z} = \begin{bmatrix} c\hat{\theta}_n & -s\hat{\theta}_n & 0 \\ c\hat{\alpha}_{n-1}\, s\hat{\theta}_n & c\hat{\alpha}_{n-1}\, c\hat{\theta}_n & -s\hat{\alpha}_{n-1} \\ s\hat{\alpha}_{n-1}\, s\hat{\theta}_n & s\hat{\alpha}_{n-1}\, c\hat{\theta}_n & c\hat{\alpha}_{n-1} \end{bmatrix} \tag{199}$$

or if the link length is on the *y*-axis and the joint at the proximal end, then

$$ \hat{T}_{YZ} = \hat{Z}\hat{Y} = \begin{bmatrix} c\hat{\alpha}_{n1}\, c\hat{\theta}_n & -s\hat{\theta}_n & s\hat{\alpha}_n\, c\hat{\theta}_n \\ c\hat{\alpha}_n\, s\hat{\theta}_n & c\hat{\theta}_n & s\hat{\alpha}_n\, s\hat{\theta}_n \\ -s\hat{\alpha}_{n-1} & 0 & c\hat{\alpha}_n \end{bmatrix} \tag{200}$$

Any of these formulations can be used to properly predict the reality of how the links of a mechanism are situated relative to each other. The resulting equations will appear different and so it is important to understand to what values the parameters and variables refer.

Exercises - Matrix Modeling of Joints and Links

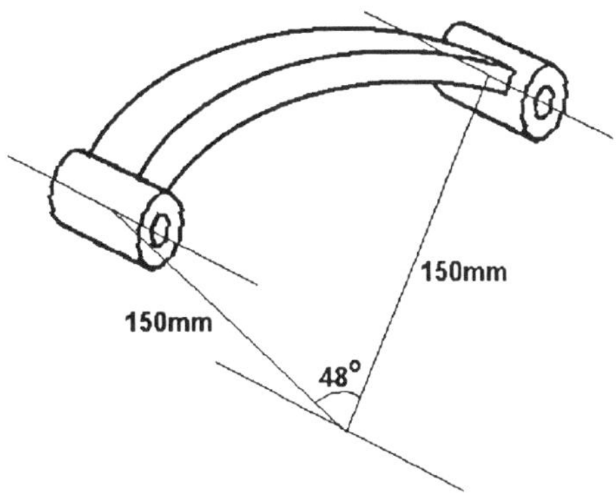

Figure 17

1. What are the length and twist of the link shown in Figure 17?

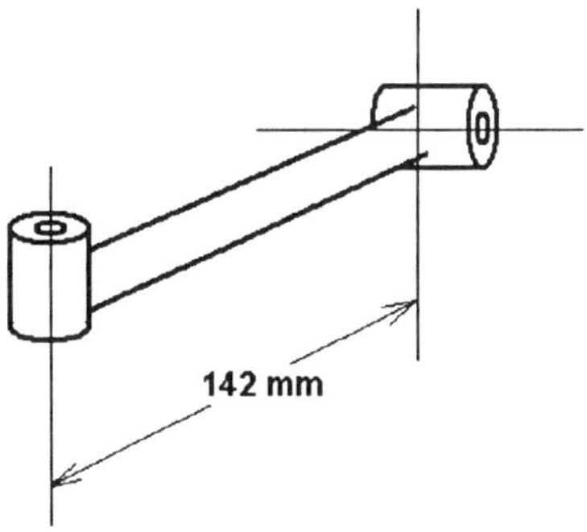

Figure 18

2. What are the length and twist of the link shown in Figure 18?

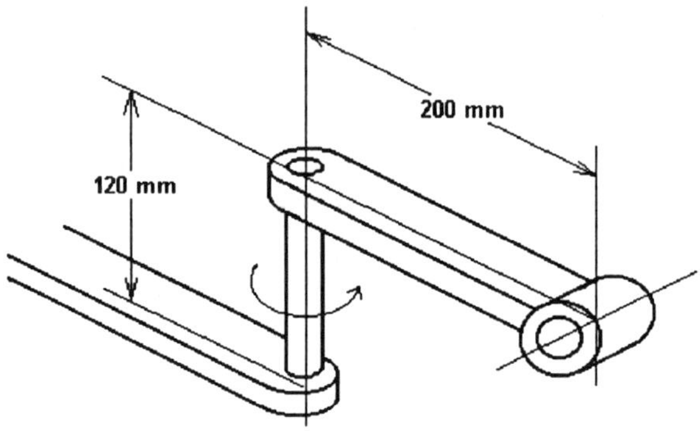

Figure 19

3. What are the parameters which describe the joint and link shown in Figure 19?

Condition of Loop Closure

The displacement analysis of a mechanism consists of determining the rotations and translations at the output and intermediate joints given the dimensions of the links and the displacements at the input joint (or input joints since sometimes there are more than one).

Suppose that a mechanism has four links which may be represented by the joint-link modeling matrices ${}^{1}_{2}\hat{M}$, ${}^{2}_{3}\hat{M}$, ${}^{3}_{4}\hat{M}$, ${}^{4}_{1}\hat{M}$. If coordinate transformations are made from the input link to the coupler link, then to the output link, to the ground link and to the input link, the total effect of the transformations is a return to the initial coordinate system. This is known as the "condition of loop closure" and is expressed mathematically as

$$ {}^{1}_{2}\hat{M}\,{}^{2}_{3}\hat{M}\,{}^{3}_{4}\hat{M}\,{}^{4}_{1}\hat{M} = \hat{I} \tag{201} $$

where symbol $\hat{I}$ represents the dual identity matrix. Analogous equations can be written for mechanisms featuring closed chains of more than four links. The unknown displacements of the mechanism can be found by solution of this equation.

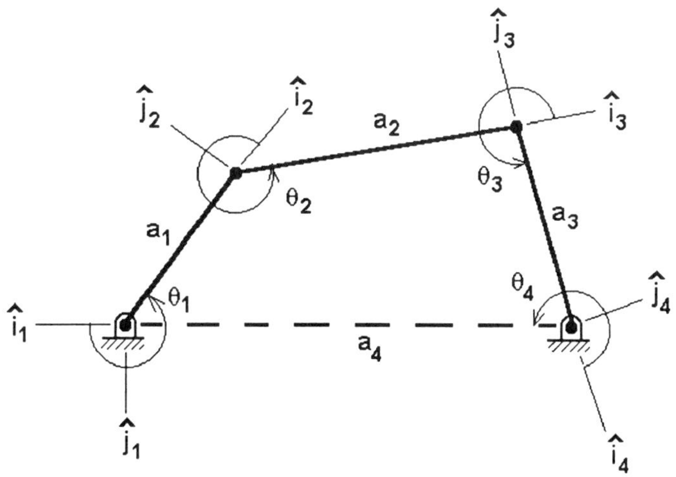

Figure 20 Displacements of the planar four-bar mechanism.

Planar Four-Bar Mechanism

Consider a planar four-bar mechanism, as shown in Figure 20, which is a mechanism of four links, considered to act within a plane, connected by revolute joints. It is a one degree-of-freedom mechanism and the input variable is the orientation of the input crank, rotation θ_1. The input crank has a length a_1, the coupler link has length a_2. the output crank has length a_3 and the stationary ground link has a length a_4. It is desired to determine the output rotation θ_4 and the intermediate rotations θ_2 and θ_3.

An organized way in which to analyze the displacements in this mechanism is to make up a table of the Denavit-Hartenberg parameters, establish the joint-link transforms, substitute into the condition-of-loop-closure equation and then manipulate that expression to obtain the desired results.

Table 1 - Planar Four-Bar Mechanism

i	name	θ_i	s_i	α_i	a_i
1	input	θ_1	0	0°	a_1
2	coupler	θ_2	0	0°	a_2
3	output	θ_3	0	0°	a_3
4	ground	θ_4	0	0°	a_4

We can make up a table of the Denavit-Hartenberg parameters for this mechanism considering the link twists and lengths and the permissible motions at the joints. As shown in Table 1, the link twists α_i are all set to zero since the joint axes are parallel to each other. The joint offsets s_i are set to zero since the mechanism is considered to be planar. If overlapping links are to be considered, then the appropriate offsets would be set to constant values depending on the size of the offsets.

For the particular values shown in the Denavit-Hartenberg parameter table, it is observed that

$$\mathrm{s}\hat{\theta}_n = \mathrm{s}\theta_n + \varepsilon s_n \mathrm{c}\theta_n \quad \rightarrow \quad \mathrm{s}\hat{\theta}_n = \mathrm{s}\theta_n$$

$$\mathrm{c}\hat{\theta}_n = \mathrm{c}\theta_n - \varepsilon s_n \mathrm{s}\theta_n \quad \rightarrow \quad \mathrm{c}\hat{\theta}_n = \mathrm{c}\theta_n$$

$$\mathrm{s}\hat{\alpha}_n = \mathrm{s}\alpha_n + \varepsilon a_n \mathrm{c}\alpha_n \quad \rightarrow \quad \mathrm{s}\hat{\alpha}_n = \varepsilon a_n$$

$$\mathrm{c}\hat{\alpha}_n = \mathrm{c}\alpha_n - \varepsilon a_n \mathrm{s}\alpha_n \quad \rightarrow \quad \mathrm{c}\hat{\alpha}_n = 1$$

Substituting these results into the joint-link modeling transformation matrices, we obtain

$$ {}_{n+1}^{\;\;n}\hat{\mathrm{M}} = \begin{bmatrix} c\theta_n & -s\theta_n & \varepsilon a_n s\theta_n \\ s\theta_n & c\theta_n & -\varepsilon a_n c\theta_n \\ 0 & \varepsilon a_n & 1 \end{bmatrix} \tag{202} $$

These four matrices are substituted into the condition of loop closure:

$$ {}_2^1\hat{\mathrm{M}}\,{}_3^2\hat{\mathrm{M}}\,{}_4^3\hat{\mathrm{M}}\,{}_1^4\hat{\mathrm{M}} = \hat{\mathrm{I}} \tag{203} $$

It would seem that the thing to do is to expand this equation and then equate the elements as necessary to obtain expressions for the unknown displacements θ_2, θ_3 and θ_4. In actual practice, it is advantageous to first manipulate this equation so that the intermediate-joint displacements are on one side and the input- and output-joint displacements are on the other.

$$ {}_3^2\hat{\mathrm{M}}\,{}_4^3\hat{\mathrm{M}}\,{}_1^4\hat{\mathrm{M}} = {}_2^1\hat{\mathrm{M}}^{-1} \tag{204} $$

$$ {}_3^2\hat{\mathrm{M}}\,{}_4^3\hat{\mathrm{M}} = {}_2^1\hat{\mathrm{M}}^{-1}\,{}_1^4\hat{\mathrm{M}}^{-1} \tag{205} $$

$$ {}_3^2\hat{\mathrm{M}}\,{}_4^3\hat{\mathrm{M}} = {}_2^1\hat{\mathrm{M}}^{T}\,{}_1^4\hat{\mathrm{M}}^{T} \tag{206} $$

$$ \begin{bmatrix} c\theta_2 & -s\theta_2 & \varepsilon a_2 s\theta_2 \\ s\theta_2 & c\theta_2 & -\varepsilon a_2 c\theta_2 \\ 0 & \varepsilon a_2 & 1 \end{bmatrix} \begin{bmatrix} c\theta_3 & -s\theta_3 & \varepsilon a_3 s\theta_3 \\ s\theta_3 & c\theta_3 & -\varepsilon a_3 c\theta_3 \\ 0 & \varepsilon a_3 & 1 \end{bmatrix} = \begin{bmatrix} c\theta_1 & s\theta_1 & 0 \\ -s\theta_1 & c\theta_1 & \varepsilon a_1 \\ \varepsilon a_1 s\theta_1 & -\varepsilon a_1 c\theta_1 & 1 \end{bmatrix} \begin{bmatrix} c\theta_4 & s\theta_4 & 0 \\ -s\theta_4 & c\theta_4 & \varepsilon a_4 \\ \varepsilon a_4 s\theta_4 & -\varepsilon a_4 c\theta_4 & 1 \end{bmatrix} \tag{207} $$

$$ \begin{bmatrix} c\theta_2 c\theta_3 - s\theta_2 s\theta_3 & -c\theta_2 s\theta_3 - s\theta_2 c\theta_3 & \varepsilon a_3 c\theta_2 s\theta_3 + \varepsilon a_3 s\theta_2 c\theta_3 + \varepsilon a_2 s\theta_2 \\ s\theta_2 c\theta_3 + c\theta_2 s\theta_3 & -s\theta_2 s\theta_3 + c\theta_2 c\theta_3 & \varepsilon a_3 s\theta_2 s\theta_3 - \varepsilon a_3 c\theta_2 c\theta_3 - \varepsilon a_2 c\theta_2 \\ \varepsilon a_2 s\theta_3 & \varepsilon a_2 c\theta_3 + \varepsilon a_3 & 1 \end{bmatrix} = \begin{bmatrix} c\theta_1 c\theta_4 - s\theta_1 s\theta_4 & c\theta_1 s\theta_4 + s\theta_1 c\theta_4 & \varepsilon a_4 s\theta_1 \\ -s\theta_1 c\theta_4 - c\theta_1 s\theta_4 & -s\theta_1 c\theta_4 - c\theta_1 s\theta_4 & \varepsilon a_4 c\theta_1 \\ \varepsilon a_1 s\theta_1 c\theta_4 + \varepsilon a_1 c\theta_1 s\theta_4 + \varepsilon a_4 s\theta_4 & \varepsilon a_1 s\theta_1 s\theta_4 - \varepsilon a_1 c\theta_1 c\theta_4 - \varepsilon a_4 c\theta_4 & 1 \end{bmatrix} \tag{208} $$

Equating the 3-1 elements,

$$ \varepsilon a_2 s\theta_3 = \varepsilon a_1 s\theta_1 c\theta_4 + \varepsilon a_1 c\theta_1 s\theta_4 + \varepsilon a_4 s\theta_4 \tag{209} $$

which can be manipulated to

$$s\theta_3 = \frac{a_1(s\theta_1\, c\theta_4 + c\theta_1\, s\theta_4) + a_4\, s\theta_4}{a_2} \tag{210}$$

Equating the 3-2 elements,

$$\varepsilon a_2\, c\theta_3 + \varepsilon a_3 = \varepsilon a_1\, s\theta_1\, s\theta_4 - \varepsilon a_1\, c\theta_1\, c\theta_4 - \varepsilon a_4\, c\theta_4 \tag{211}$$

which is put in the form

$$c\theta_3 = \frac{a_1(s\theta_1\, s\theta_4 - c\theta_1\, c\theta_4) - a_4\, c\theta_4 - a_3}{a_2} \tag{212}$$

Squaring both sides of the equations for the sine and cosine of intermediate angle θ_3 and then simplifying,

$$-a_1^2 + a_2^2 - a_3^2 - a_4^2 - 2a_1 a_4\, c\theta_1 - 2a_3 a_4\, c\theta_4 = 2a_1 a_3\, c(\theta_1 + \theta_4) \tag{213}$$

This result is the famous Freudenstein's equation. The differences in defining the angles should be considered when comparing to other formulations of Freudenstein's equation.

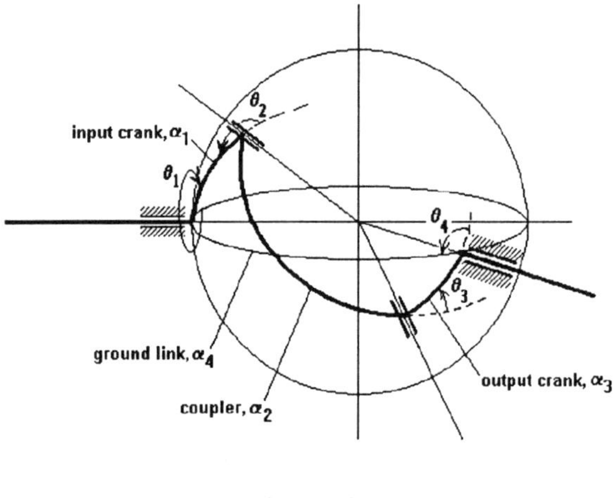

Figure 21 Displacements of the spherical four-bar mechanism.

Spherical Four-Bar Mechanism

A spherical mechanism is a spatial linkage whose joints have axes that intersect at a common point. Therefore, the shortest distance between the joint axes of each link $a_i = 0$ for all i. The links may be thought of as arcs floating on the surface of a sphere. The angular dimension of each link is its twist angle α_i. Such a mechanism is shown in Figure 21.

Table 2 - Spherical Four-Bar Mechanism

i	θ_i	s_i	α_i	a_i
1	θ_1	0	α_1	0
2	θ_2	0	α_2	0
3	θ_3	0	α_3	0
4	θ_4	0	α_4	0

The joints in a spherical four-bar mechanism are all revolute, therefore joint offset $s_i = 0$ for all i. We can set up the Denavit-Hartenberg parameters for this mechanism as shown in Table 2.

Using the values in the table, the joint link matrices can readily be written:

$$ {}^{i}_{i+1}\hat{\mathrm{M}} = \begin{bmatrix} c\theta_i & -c\alpha_i s\theta_i & s\alpha_i s\theta_i \\ s\theta_i & c\alpha_i c\theta_i & -s\alpha_i c\theta_i \\ 0 & s\alpha_i & c\alpha_i \end{bmatrix} \tag{214} $$

The four matrices represented by this equation are substituted into the condition of loop closure:

$$ {}^{1}_{2}\hat{\mathrm{M}}\,{}^{2}_{3}\hat{\mathrm{M}}\,{}^{3}_{4}\hat{\mathrm{M}}\,{}^{4}_{1}\hat{\mathrm{M}} = \hat{\mathrm{I}} \tag{215} $$

To solve for the output-joint displacement θ_4 intermediate-joint displacements θ_2 and θ_3 this equation is manipulated to the form

$$ {}^{2}_{3}\hat{\mathrm{M}}\,{}^{3}_{4}\hat{\mathrm{M}} = {}^{1}_{2}\hat{\mathrm{M}}^{\mathrm{T}}\,{}^{4}_{1}\hat{\mathrm{M}}^{\mathrm{T}} \tag{216} $$

and the expressions for the joint-link matrices substituted and then the solutions developed.

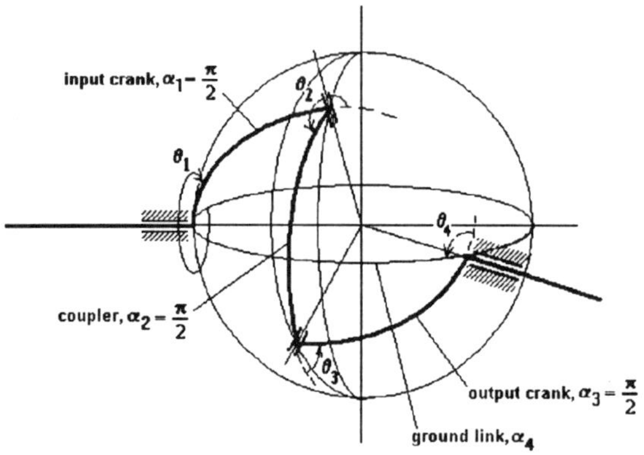

Figure 22 Displacements of the Cardan joint.

Cardan Joint

The Cardan joint, often referred to as the Hooke's coupling or the universal joint, is a form of spherical four-bar mechanism widely used as a shaft coupling in automotive and machinery drive lines.

Table 3 - The Cardan Joint

i	θ_i	s_i	α_i	a_i
1	θ_1	0	90°	0
2	θ_2	0	90°	0
3	θ_3	0	90°	0
4	θ_4	0	α_4	0

In the Cardan joint, the angular dimensions of the moving links are $\alpha_1 = \alpha_2 = \alpha_3 = 90°$, and the ground-link angle α_4 has the value 180° in the "straight through" configuration. The practical limit on the ground-link angle is generally considered to be 150°.

The Denavit-Hartenberg parameters are shown in Table 4. The table allows ready substitution of the parameters into the joint-link modeling matrices.

The equation developed by considering the condition of loop closure,

$$ {}^{1}_{2}\hat{M}\,{}^{2}_{3}\hat{M}\,{}^{3}_{4}\hat{M}\,{}^{4}_{1}\hat{M} = \hat{I} \tag{217}$$

is easier to solve when manipulated to the form

$$ {}^{2}_{3}\hat{M}\,{}^{3}_{4}\hat{M} = {}^{1}_{2}\hat{M}^{T}\,{}^{4}_{1}\hat{M}^{T} \tag{218}$$

so that the input and output variables, θ_1 and θ_4, are on the right-hand side, while the intermediate variables θ_2 and θ_3 appear on the left-hand side.

$$\begin{bmatrix} c\theta_2 & 0 & s\theta_2 \\ s\theta_2 & 0 & -c\theta_2 \\ 0 & 1 & 0 \end{bmatrix}\begin{bmatrix} c\theta_3 & 0 & s\theta_3 \\ s\theta_3 & 0 & -c\theta_3 \\ 0 & 1 & 0 \end{bmatrix} = \begin{bmatrix} c\theta_1 & s\theta_1 & 0 \\ 0 & 0 & 1 \\ s\theta_1 & -c\theta_1 & 0 \end{bmatrix}\begin{bmatrix} c\theta_4 & s\theta_4 & 0 \\ -c\alpha_4\,s\theta_4 & c\alpha_4\,c\theta_4 & s\alpha_4 \\ s\alpha_4\,s\theta_4 & -s\alpha_4\,c\theta_4 & c\alpha_4 \end{bmatrix} \tag{219}$$

$$\begin{bmatrix} c\theta_2\,c\theta_3 & s\theta_2 & c\theta_2\,s\theta_3 \\ s\theta_2\,c\theta_3 & -c\theta_2 & s\theta_2\,s\theta_3 \\ s\theta_3 & 0 & -c\theta_3 \end{bmatrix} = \begin{bmatrix} c\theta_1\,c\theta_4 - c\alpha_4\,s\theta_1\,s\theta_4 & c\theta_1\,s\theta_4 + c\alpha_4\,s\theta_1\,c\theta_4 & s\alpha_4\,s\theta_1 \\ s\alpha_4\,s\theta_4 & -s\alpha_4\,c\theta_4 & c\alpha_4 \\ s\theta_1\,c\theta_4 + c\alpha_4\,c\theta_1\,s\theta_4 & s\theta_1\,s\theta_4 - c\alpha_4\,c\theta_1\,c\theta_4 & -s\alpha_4\,c\theta_1 \end{bmatrix} \tag{220}$$

Equating the 3-2 elements of this equation, we obtain

$$0 = s\theta_1\,s\theta_4 - c\alpha_4\,c\theta_1\,c\theta_4 \tag{221}$$

which rearranges to

$$\cos\alpha_4 = \tan\theta_1\,\tan\theta_4 \tag{222}$$

This is the input-output relationship for the Cardan joint. Allowing for how the angles are measured, this is equivalent to the result obtained by traditional analyses.

We can also solve for the intermediate-joint displacements.

To solve for the rotation θ_2, the 1-2 and 2-2 terms are equated,

$$s\theta_2 = c\theta_1\,s\theta_4 + c\alpha_4\,s\theta_1\,c\theta_4 \tag{223}$$

$$-c\theta_2 = -s\alpha_4\,c\theta_4 \tag{224}$$

To solve for the rotation θ_3, the 3-1 and 3-3 terms are equated,

$$s\theta_3 = s\theta_1\, c\theta_4 + c\alpha_4\, c\theta_1\, s\theta_4 \tag{225}$$

$$-c\theta_3 = -\, s\alpha_4\, c\theta_1 \tag{226}$$

It is necessary to have expressions for both the sine and cosine of θ_1 and θ_2 since either alone is ambiguous as to the quadrant of the angle. Consideration of the signs of the sine and cosine of an angle taken together indicate the proper quadrant as shown in Table 4. In various programming languages and most hand-held calculators where the arctangent of an angle always returns a value between $+^{\pi}/_2$ and $-^{\pi}/_2$ radians, the proper quadrant is indicated if both the sine and cosine of the angle are positive. Otherwise, π radians must be added to the result if the cosine of the angle is negative, and 2π radians must be added if the sine of the angle is negative and the cosine is positive.

Table 4 - Quadrant of an Angle

sine	cosine	quadrant
+	+	1
+	–	2
–	–	3
–	+	4

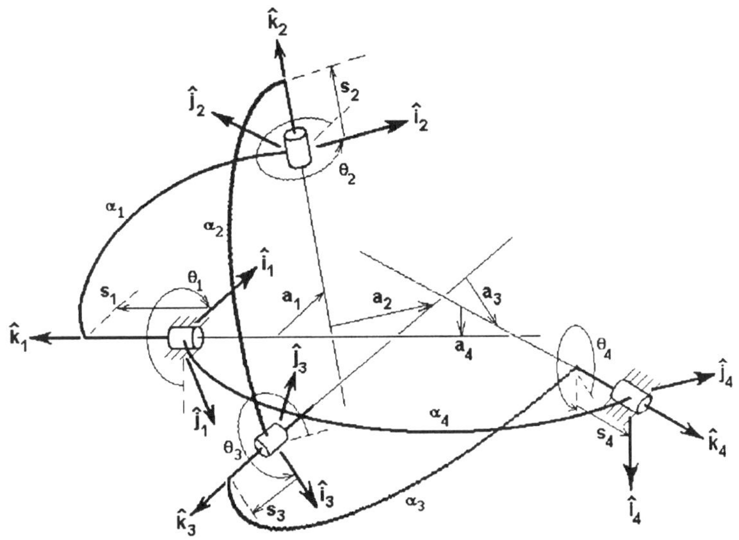

Figure 23 Displacements of a general spatial four-bar mechanism.

General Spatial Four-Bar Mechanism

The general spatial four-bar mechanism, shown in Figure 23, consists of four links with axes which are not collinear as are those of a planar mechanism and which do not all intersect as do those of a spherical mechanism. The spatial four-bar mechanism is generally considered as a one degree-of-freedom RC^3 linkage. For generality, we will consider it as a two degree-of-freedom C^4 linkage.

It is desired to determine the displacements at the intermediate and output joints as functions of the displacements at the input joint. Thus the unknowns are the rotations θ_2, θ_3 and θ_4 and the translations s_2, s_3 and s_4. The input variables are the rotation and translation at the input joint, θ_1 and s_1. The dimensions of the links, dual angles $\hat{\alpha}_1, \hat{\alpha}_2, \hat{\alpha}_3, \hat{\alpha}_4$ are given.

Table 5 - The General Spatial Four-Bar Mechanism

i	θ_i	s_i	α_i	a_i
1	θ_1	s_1	α_1	a_1
2	θ_2	s_2	α_2	a_2
3	θ_3	s_3	α_3	a_3
4	θ_4	s_4	α_4	a_4

The condition of loop closure for a four-link chain is manipulated from

$$ {}^{1}_{2}\hat{M}\,{}^{2}_{3}\hat{M}\,{}^{3}_{4}\hat{M}\,{}^{4}_{1}\hat{M} = \hat{I} \tag{227} $$

into the form

$$ {}^{2}_{3}\hat{M}\,{}^{3}_{4}\hat{M} = {}^{1}_{2}\hat{M}^{T}\,{}^{4}_{1}\hat{M}^{T} \tag{228} $$

Substituting the values of the Denavit-Hartenberg parameters from Table 5 into the joint-link modeling matrices in the condition-of-loop closure equation, then

$$ \begin{bmatrix} c\hat{\theta}_2 & -c\hat{\alpha}_2\, s\hat{\theta}_2 & s\hat{\alpha}_2\, s\hat{\theta}_2 \\ s\hat{\theta}_2 & c\hat{\alpha}_2\, c\hat{\theta}_2 & -s\hat{\alpha}_2\, c\hat{\theta}_2 \\ 0 & s\hat{\alpha}_2 & c\hat{\alpha}_2 \end{bmatrix} \begin{bmatrix} c\hat{\theta}_3 & -c\hat{\alpha}_3\, s\hat{\theta}_3 & s\hat{\alpha}_3\, s\hat{\theta}_3 \\ s\hat{\theta}_3 & c\hat{\alpha}_3\, c\hat{\theta}_3 & -s\hat{\alpha}_3\, c\hat{\theta}_3 \\ 0 & s\hat{\alpha}_3 & c\hat{\alpha}_3 \end{bmatrix} = \begin{bmatrix} c\hat{\theta}_1 & s\hat{\theta}_1 & 0 \\ -c\hat{\alpha}_1\, s\hat{\theta}_1 & c\hat{\alpha}_1\, c\hat{\theta}_1 & s\hat{\alpha}_1 \\ s\hat{\alpha}_1\, s\hat{\theta}_1 & -s\hat{\alpha}_1\, c\hat{\theta}_1 & c\hat{\alpha}_1 \end{bmatrix} \begin{bmatrix} c\hat{\theta}_4 & s\hat{\theta}_4 & 0 \\ -c\hat{\alpha}_4\, s\hat{\theta}_4 & c\hat{\alpha}_4\, c\hat{\theta}_4 & s\hat{\alpha}_4 \\ s\hat{\alpha}_4\, s\hat{\theta}_4 & -s\hat{\alpha}_4\, c\hat{\theta}_4 & c\hat{\alpha}_4 \end{bmatrix} \tag{229} $$

$$\begin{bmatrix} c\hat{\theta}_2\, c\hat{\theta}_3 - c\hat{\alpha}_2\, s\hat{\theta}_2\, s\hat{\theta}_3 & \begin{matrix} s\hat{\alpha}_2\, s\hat{\alpha}_3\, s\hat{\theta}_2 \\ -c\hat{\alpha}_3\left(c\hat{\theta}_2\, s\hat{\theta}_3 + c\hat{\alpha}_2\, s\hat{\theta}_2\, c\hat{\theta}_3\right) \end{matrix} & \begin{matrix} s\hat{\alpha}_2\, c\hat{\alpha}_3\, s\hat{\theta}_2 \\ +s\hat{\alpha}_3\left(c\hat{\theta}_2\, s\hat{\theta}_3 + c\hat{\alpha}_2\, s\hat{\theta}_2\, c\hat{\theta}_3\right) \end{matrix} \\ s\hat{\theta}_2\, c\hat{\theta}_3 + c\hat{\alpha}_2\, c\hat{\theta}_2\, s\hat{\theta}_3 & \begin{matrix} -s\hat{\alpha}_2\, s\hat{\alpha}_3\, c\hat{\theta}_2 \\ -c\hat{\alpha}_3\left(s\hat{\theta}_2\, s\hat{\theta}_3 - c\hat{\alpha}_2\, c\hat{\theta}_2\, c\hat{\theta}_3\right) \end{matrix} & \begin{matrix} -s\hat{\alpha}_2\, c\hat{\alpha}_3\, c\hat{\theta}_2 \\ -s\hat{\alpha}_3\left(s\hat{\theta}_2\, s\hat{\theta}_3 - c\hat{\alpha}_2\, c\hat{\theta}_2\, c\hat{\theta}_3\right) \end{matrix} \\ s\hat{\alpha}_2\, s\hat{\theta}_3 & c\hat{\alpha}_2\, s\hat{\alpha}_3 + s\hat{\alpha}_2\, c\hat{\alpha}_3\, c\hat{\theta}_3 & c\hat{\alpha}_2\, c\hat{\alpha}_3 - s\hat{\alpha}_2\, s\hat{\alpha}_3\, c\hat{\theta}_3 \end{bmatrix} \quad (230)$$

$$= \begin{bmatrix} c\hat{\theta}_1\, c\hat{\theta}_4 - c\hat{\alpha}_4\, s\hat{\theta}_1\, s\hat{\theta}_4 & c\hat{\theta}_1\, s\hat{\theta}_4 + c\hat{\alpha}_4\, s\hat{\theta}_1\, c\hat{\theta}_4 & s\hat{\alpha}_4\, s\hat{\theta}_1 \\ \begin{matrix} s\hat{\alpha}_1\, s\hat{\alpha}_4\, s\hat{\theta}_1 \\ -c\hat{\alpha}_1\left(s\hat{\theta}_1\, c\hat{\theta}_4 - c\hat{\alpha}_4\, c\hat{\theta}_1\, s\hat{\theta}_4\right) \end{matrix} & \begin{matrix} -s\hat{\alpha}_1\, s\hat{\alpha}_4\, c\hat{\theta}_1 \\ -c\hat{\alpha}_1\left(s\hat{\theta}_1\, s\hat{\theta}_4 - c\hat{\alpha}_4\, c\hat{\theta}_1\, c\hat{\theta}_4\right) \end{matrix} & s\hat{\alpha}_1\, c\hat{\alpha}_4 + c\hat{\alpha}_1\, s\hat{\alpha}_4\, c\hat{\theta}_1 \\ \begin{matrix} c\hat{\alpha}_1\, s\hat{\alpha}_4\, s\hat{\theta}_4 \\ +s\hat{\alpha}_1\left(s\hat{\theta}_1\, c\hat{\theta}_4 + c\hat{\alpha}_4\, c\hat{\theta}_1\, s\hat{\theta}_4\right) \end{matrix} & \begin{matrix} -c\hat{\alpha}_1\, s\hat{\alpha}_4\, c\hat{\theta}_4 \\ +s\hat{\alpha}_1\left(s\hat{\theta}_1\, s\hat{\theta}_4 - c\hat{\alpha}_4\, c\hat{\theta}_1\, c\hat{\theta}_4\right) \end{matrix} & c\hat{\alpha}_1\, c\hat{\alpha}_4 - s\hat{\alpha}_1\, s\hat{\alpha}_4\, c\hat{\theta}_1 \end{bmatrix}$$

Output Angular Displacement θ_4

Equating the 3-2 elements of eq.(230)

$$c\hat{\alpha}_2\, s\hat{\alpha}_3 + s\hat{\alpha}_2\, c\hat{\alpha}_3\, c\hat{\theta}_3 = -c\hat{\alpha}_1\, s\hat{\alpha}_4\, c\hat{\theta}_4 + s\hat{\alpha}_1\left(s\hat{\theta}_1\, s\hat{\theta}_4 - c\hat{\alpha}_4\, c\hat{\theta}_1\, c\hat{\theta}_4\right) \quad (231)$$

Equating the 3-3 elements of eq.(230)

$$c\hat{\alpha}_2\, c\hat{\alpha}_3 - s\hat{\alpha}_2\, s\hat{\alpha}_3\, c\hat{\theta}_3 = c\hat{\alpha}_1\, c\hat{\alpha}_4 - s\hat{\alpha}_1\, s\hat{\alpha}_4\, c\hat{\theta}_1 \quad (232)$$

Multiplying eq.(231) by $s\alpha_3$ and eq.(232) by $c\hat{\alpha}_3$ and adding and simplifying

$$c\hat{\alpha}_2 = -c\hat{\alpha}_1\, s\hat{\alpha}_3\, s\hat{\alpha}_4\, c\hat{\theta}_4 + s\hat{\alpha}_1\, s\hat{\alpha}_3\left(s\hat{\theta}_1\, s\hat{\theta}_4 - c\hat{\alpha}_4\, c\hat{\theta}_1\, c\hat{\theta}_4\right) + c\hat{\alpha}_1\, c\hat{\alpha}_3\, c\hat{\alpha}_4 - s\hat{\alpha}_1\, c\hat{\alpha}_3\, s\hat{\alpha}_4\, c\hat{\theta}_1 \quad (233)$$

This is rearranged to the form

$$0 = \hat{A}\sin\hat{\theta}_4 + \hat{B}\cos\hat{\theta}_4 + \hat{C} \quad (234)$$

where

$$\hat{A} = s\hat{\alpha}_1\, s\hat{\alpha}_3\, s\hat{\theta}_1 \quad (235)$$

$$\hat{B} = -s\hat{\alpha}_3\left(c\hat{\alpha}_1\, s\hat{\alpha}_4 + s\hat{\alpha}_1\, c\hat{\alpha}_4\, c\hat{\theta}_1\right) \quad (236)$$

$$\hat{C} = -c\hat{\alpha}_2 + c\hat{\alpha}_3\left(c\hat{\alpha}_1\, c\hat{\alpha}_4 - s\hat{\alpha}_1\, s\hat{\alpha}_4\, c\hat{\theta}_1\right) \quad (237)$$

The primary component of eq.(234) is

$$
\begin{aligned}
0 = \\
&(s\alpha_1\, s\alpha_3\, s\theta_1)\, s\theta_4 \\
&-\left[s\alpha_3\left(c\alpha_1\, s\alpha_4 + s\alpha_1\, c\alpha_4\, c\theta_1\right)\right] c\theta_4 \\
&+\left[c\alpha_3\left(c\alpha_1\, c\alpha_4 - s\alpha_1\, s\alpha_4\, c\theta_1\right) - c\alpha_2\right]
\end{aligned}
\tag{238}
$$

This equation shows how the output displacement angle, θ_4, is related to the input displacement angle, θ_1, and also to the angular dimensions of the links.

To solve explicitly for angle θ_4, the half-tangent relationships are employed. These are

$$\sin\theta_4 = \frac{2t}{1+t^2} \tag{239}$$

$$\cos\theta_4 = \frac{1-t^2}{1+t^2} \tag{240}$$

where

$$t = \tan\frac{\theta_4}{2} \tag{241}$$

So, if we let

$$D = s\alpha_1\, s\alpha_3\, s\theta_1 \tag{242}$$

$$E = -\, s\alpha_3\left(c\alpha_1\, s\alpha_4 + s\alpha_1\, c\alpha_4\, c\theta_1\right) \tag{243}$$

$$F = c\alpha_3\left(c\alpha_1\, c\alpha_4 - s\alpha_1\, s\alpha_4\, c\theta_1\right) - c\alpha_2 \tag{244}$$

and substitute these equations with the half-tangent relationships into eq.(238), we obtain

$$(F-E)t^2 + 2Dt + (E+F) = 0 \tag{245}$$

The solution of this equation is

$$t = \frac{-D \pm \sqrt{D^2 + E^2 - F^2}}{(F-E)} \tag{246}$$

where

$$\theta_4 = 2\tan^{-1} t \tag{247}$$

Thus the output displacement angle, θ_4, can be found by substituting the given input displacement angle, θ_1, and the angular dimensions of the links into eqs.(242,243,244) which give the equations to be substituted into eq.(246), the solution of which is substituted into eq.(247) to yield the desired output angular displacement.

It should be noted that eq.(246) has two solutions. The physical reality corresponding to these two solutions is shown in Figure 24. The positive root indicates the value of angle θ_4 when the mechanism is assembled so that the input link leads the output link. If the mechanism is assembled so that the output link leads the input link, then angle θ_4 is found by using the negative root of eq.(246).

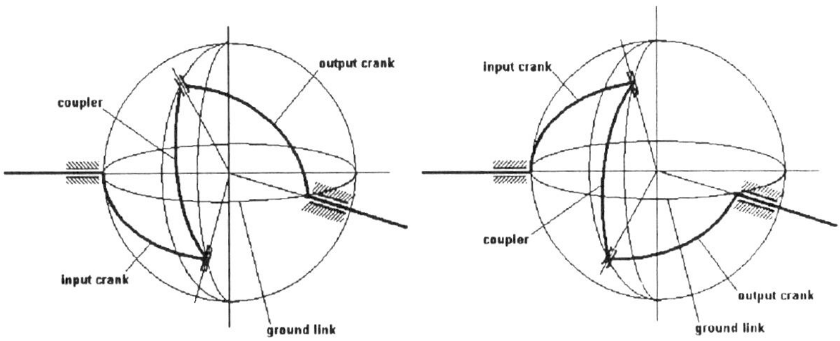

Figure 24 Cardan joints assembled to show how in a spatial four-bar mechanism the input link can lead the output link (left) or output link can lead the input link (right).

Translational Displacement s_4

The dual component of eq.(234) is

$$
\begin{aligned}
0 = {} & a_1\{c\alpha_1[s\alpha_3(s\theta_1\, s\theta_4 - c\alpha_4\, c\theta_1\, c\theta_4) - c\alpha_3\, s\alpha_4\, c\theta_1] + s\alpha_1(c\alpha_3\, c\alpha_4 - s\alpha_3\, s\alpha_4\, c\theta_4)\} \\
& + a_2\, s\alpha_2 \\
& + a_3\{s\alpha_1[c\alpha_3(s\theta_1\, s\theta_4 - c\alpha_4\, c\theta_1\, c\theta_4) + s\alpha_3\, s\alpha_4\, c\theta_1] - c\alpha_1(s\alpha_3\, c\alpha_4 + c\alpha_3\, s\alpha_4\, c\theta_4)\} \\
& + a_4\{-s\alpha_1\, c\theta_1(c\alpha_3\, c\alpha_4 - s\alpha_3\, s\alpha_4\, c\theta_4) + c\alpha_1(c\alpha_3\, s\alpha_4 + s\alpha_3\, c\alpha_4\, c\theta_4)\} \\
& + s_1\{s\alpha_1[s\alpha_3(c\theta_1\, s\theta_4 + c\alpha_4\, s\theta_1\, c\theta_4) + c\alpha_3\, s\alpha_4\, s\theta_4]\} \\
& + s_4\{s\alpha_3[s\alpha_1(s\theta_1\, c\theta_4 + c\alpha_4\, c\theta_1\, s\theta_4) + c\alpha_3\, s\alpha_4\, s\theta_4]\}
\end{aligned}
\tag{248}
$$

This equation gives the displacement s_4, the sliding at the output joint as a function of input angle θ_4 and input translation s_1 and also the angular dimensions of the links and the offsets of the joint axes.

Intermediate Rotation θ_3 and Translation s_3

Equating the 3-1 terms of eq.(230),

$$s\hat{\alpha}_2 s\hat{\theta}_3 = c\hat{\alpha}_1 s\hat{\alpha}_4 s\hat{\theta}_4 + s\hat{\alpha}_1\left(s\hat{\theta}_1 c\hat{\theta}_4 + c\hat{\alpha}_4 c\hat{\theta}_1 s\hat{\theta}_4\right) \tag{249}$$

The primary portion of eq.(249) can be solved to obtain $s\theta_3$

$$s\theta_3 = \frac{c\alpha_1 s\alpha_4 s\theta_4 + s\alpha_1\left(s\theta_1 c\theta_4 + c\alpha_4 c\theta_1 s\theta_4\right)}{s\alpha_2} \tag{250}$$

Equating the 3-3 terms of eq.(230),

$$c\hat{\alpha}_2 c\hat{\alpha}_3 - s\hat{\alpha}_2 s\hat{\alpha}_3 c\hat{\theta}_3 = c\hat{\alpha}_1 c\hat{\alpha}_4 - s\hat{\alpha}_1 s\hat{\alpha}_4 c\hat{\theta}_1 \tag{251}$$

The primary component of eq.(251) can be solved to obtain $c\theta_3$

$$c\theta_3 = \frac{-c\alpha_1 c\alpha_4 + c\alpha_2 c\alpha_3 + s\alpha_1 s\alpha_4 c\theta_1}{s\alpha_2 s\alpha_3} \tag{252}$$

The dual component of eq.(251) is

$$\begin{aligned}
&s_3\, s\alpha_2\, c\theta_3 \\
&= \\
&a_1\left[c\alpha_1\left(s\theta_1\, c\theta_4 + c\alpha_4\, c\theta_1\, s\theta_4\right) - s\alpha_1\, s\alpha_4\, s\theta_4\right] \\
&- a_2\, c\alpha_2\, s\theta_3 \\
&+ a_4\left[s\theta_4\left(c\alpha_1\, c\alpha_4 - s\alpha_1\, s\alpha_4\, c\theta_1\right)\right] \\
&+ s_1\left[s\alpha_1\left(c\theta_1\, c\theta_4 - c\alpha_4\, s\theta_1\, s\theta_4\right)\right] \\
&+ s_4\left[c\alpha_1\, s\alpha_4\, c\theta_4 - s\alpha_1\left(s\theta_1\, s\theta_4 - c\alpha_4\, c\theta_1\, c\theta_4\right)\right]
\end{aligned} \tag{253}$$

Eq.(253) gives the axial sliding s_3 as a function of rotation angles θ_1, θ_3 and θ_4, the axial sliding components s_1 and s_4, and the dimensions of the mechanism.

Intermediate Rotation θ_2 and Translation s_2

Equating the 1-2 elements of eq.(230),

$$s\hat{\alpha}_2 s\hat{\alpha}_3 s\hat{\theta}_2 - c\hat{\alpha}_3\left(c\hat{\theta}_2 s\hat{\theta}_3 + c\hat{\alpha}_2 s\hat{\theta}_2 c\hat{\theta}_3\right) = c\hat{\theta}_1 s\hat{\theta}_4 + c\hat{\alpha}_4 s\hat{\theta}_1 c\hat{\theta}_4 \tag{254}$$

Equating the 1-3 elements of eq.(230)

$$s\hat{\alpha}_2\, c\hat{\alpha}_3\, s\hat{\theta}_2 + s\hat{\alpha}_3\left(c\hat{\theta}_2\, s\hat{\theta}_3 + c\hat{\alpha}_2\, s\hat{\theta}_2\, c\hat{\theta}_3\right) = s\hat{\alpha}_4\, s\hat{\theta}_1 \tag{255}$$

Adding the product of eq.(254) and $s\hat{\alpha}_3$ to the product of eq.(255) and $c\hat{\alpha}_3$,

$$s\hat{\alpha}_2 s\hat{\theta}_2 = \left(c\hat{\alpha}_3 s\hat{\alpha}_4 + s\hat{\alpha}_3 c\hat{\alpha}_4 c\hat{\theta}_4\right) s\hat{\theta}_1 + s\hat{\alpha}_3 c\hat{\theta}_1 s\hat{\theta}_4 \tag{256}$$

The primary component of eq.(256) can be solved for $s\theta_2$,

$$s\theta_2 = \frac{\left(c\alpha_3 s\alpha_4 + s\alpha_3 c\alpha_4 c\theta_4\right) s\theta_1 + s\alpha_3 c\theta_1 s\theta_4}{s\alpha_2} \tag{257}$$

To obtain an expression for $c\theta_2$, eq.(227) is manipulated to

$${}^{1}_{2}\hat{M}\, {}^{2}_{3}\hat{M} = {}^{4}_{1}\hat{M}^{T}\, {}^{3}_{4}M^{T} \tag{258}$$

which is expanded to obtain the 3-3 elements forming the equation

$$c\hat{\alpha}_1 c\hat{\alpha}_2 - s\hat{\alpha}_1 s\hat{\alpha}_2 c\hat{\theta}_2 = c\hat{\alpha}_3 c\hat{\alpha}_4 - s\hat{\alpha}_3 s\hat{\alpha}_4 c\hat{\theta}_4 \tag{259}$$

A solution can be obtained for $c\theta_2$ from the primary component of eq.(259)

$$c\theta_2 = \frac{c\alpha_1 c\alpha_2 - c\alpha_3 c\alpha_4 + s\alpha_3 s\alpha_4 c\theta_4}{s\alpha_1 s\alpha_2} \tag{260}$$

The dual component of eq.(259) is

$$\begin{gathered} s_2\, s\alpha_1\, s\alpha_2\, s\theta_2 \\ = \\ a_1\left(s\alpha_1\, c\alpha_2 + c\alpha_1\, s\alpha_2\, c\theta_2\right) \\ + a_2\left(c\alpha_1\, s\alpha_2 + s\alpha_1\, c\alpha_2\, c\theta_2\right) \\ - a_3\left(s\alpha_3\, c\alpha_4 + c\alpha_3\, s\alpha_4\, c\theta_4\right) \\ - a_4\left(c\alpha_3\, s\alpha_4 + s\alpha_3\, c\alpha_4\, c\theta_4\right) \\ + s_4\, s\alpha_3\, s\alpha_4\, s\theta_4 \end{gathered} \tag{261}$$

Eq.(261) yields axial sliding s_2 as a function of rotation angles θ_2 and θ_4, sliding s_4 and the dimensions of the mechanism.

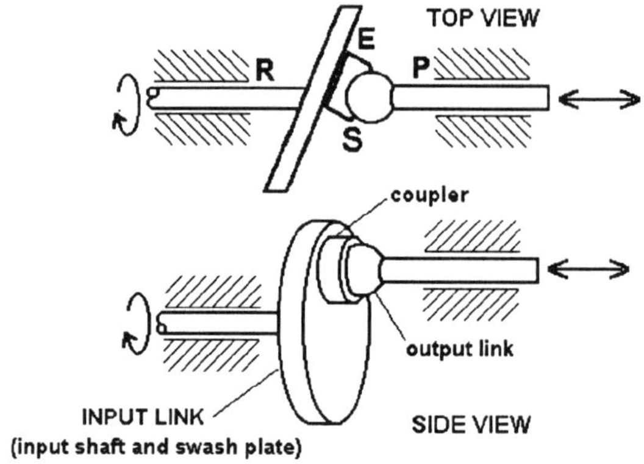

Figure 25 Swash-plate mechanism.

Swash-Plate Mechanism

The swash-plate mechanism, the RESP linkage, shown in Figure 25, features four links converting rotary to reciprocating motion and is often used where a compact arrangement is required. The axis of the input shaft can be aligned with the axis of the output slider.

Link 1 of the swash-plate mechanism is the input link and it is connected to the ground by a revolute joint which allows a rotation through input displacement θ_1. This input link is constructed in the form of a shaft on the distal end of which a plate is mounted at a twist angle α_1. The plate portion of the input link is engaged by a coupler, link 2, through a plane joint. The displacements through the plane joint are translations r_2 and s_2 and rotation θ_2. The coupler is connected to link 3, the output link, through a ball joint. Displacements within the ball joint are rotations η_3 and ζ_3. The output link is arranged to slide through translation s_4 in its prismatic joint with link 4, the ground link. The length of the ground link is distance a_4 between the axes of the shaft of the input link and the slider track of the output link. The ground link features a twist angle α_4 which has a value of 180° for the usual arrangement where the input and output joint axes are collinear. This description of the mechanism is summarized in Table 6.

Table 6 - The Swash-Plate Mechanism							
i	r_i	s_i	θ_i	η_i	ζ_i	α_i	a_i
1	-	0	θ_1	-	-	α_1	0
2	r_2	s_2	θ_2	-	-	0	0
3	-	-	0	η_3	ζ_3	0	0
4	-	s_3	0	-	-	α_4	a_4

The equation for the condition of loop closure is

$$ {}^{1}_{2}\hat{M}\,{}^{2}_{3}\hat{P}\,{}^{3}_{4}\hat{L}\,{}^{4}_{1}\hat{M} = \hat{I} \tag{262}$$

It is convenient to rearrange this equation so that all of the input and output displacements are on one side of the equal sign and all of the intermediate displacements are on the other.

$$ {}^{2}_{3}\hat{P}\,{}^{3}_{4}\hat{L} = {}^{1}_{2}\hat{M}^{T}\,{}^{4}_{1}\hat{M}^{T} \tag{263}$$

$$\begin{bmatrix} c\theta_2 & -s\theta_2 & \varepsilon s_2 \\ s\theta_2 & c\theta_2 & -\varepsilon r_2 \\ \varepsilon(r_2\, s\theta_2 - s_2\, c\theta_2) & \varepsilon(r_2\, c\theta_2 + s_2\, s\theta_2) & 1 \end{bmatrix} \begin{bmatrix} c\eta_3 & s\eta_3\, s\varsigma_3 & s\eta_3\, c\varsigma_3 \\ 0 & c\varsigma_3 & -s\varsigma_3 \\ -s\eta_3 & c\eta_3\, s\varsigma_3 & c\eta_3\, c\varsigma_3 \end{bmatrix}$$
$$= \begin{bmatrix} c\theta_1 & s\theta_1 & 0 \\ -c\alpha_1\, s\theta_1 & c\alpha_1\, c\theta_1 & s\alpha_1 \\ s\alpha_1\, s\theta_1 & -s\alpha_1\, c\theta_1 & c\alpha_1 \end{bmatrix} \begin{bmatrix} 1 & \varepsilon s_4 & 0 \\ -\varepsilon s_4\, c\alpha_4 & c\alpha_4 - \varepsilon a_4\, s\alpha_4 & s\alpha_4 + \varepsilon a_4\, c\alpha_4 \\ \varepsilon s_4\, s\alpha_4 & -s\alpha_4 - \varepsilon a_4\, c\alpha_4 & c\alpha_4 - \varepsilon a_4\, s\alpha_4 \end{bmatrix} \tag{264}$$

$$
\begin{bmatrix}
\begin{matrix} c\theta_2\, c\eta_3 \\ -\varepsilon s_2\, s\eta_3 \end{matrix} &
\begin{matrix} -s\theta_2\, c\varsigma_3 + c\theta_2\, s\eta_3\, s\varsigma_3 \\ +\varepsilon s_2\, c\eta_3\, s\varsigma_3 \end{matrix} &
\begin{matrix} s\theta_2\, s\varsigma_3 + c\theta_2\, s\eta_3\, c\varsigma_3 \\ +\varepsilon s_2\, c\eta_3\, c\varsigma_3 \end{matrix} \\[2ex]
\begin{matrix} s\theta_2\, c\eta_3 \\ +\varepsilon r_2\, s\eta_3 \end{matrix} &
\begin{matrix} c\theta_2\, c\varsigma_3 + s\theta_2\, s\eta_3\, s\varsigma_3 \\ -\varepsilon r_2\, c\eta_3\, s\varsigma_3 \end{matrix} &
\begin{matrix} -c\theta_2\, s\varsigma_3 + s\theta_2\, s\eta_3\, c\varsigma_3 \\ -\varepsilon r_2\, c\eta_3\, c\varsigma_3 \end{matrix} \\[2ex]
\begin{matrix} -s\eta_3 \\ +\varepsilon c\eta_3 (r_2\, s\theta_2 - s_2\, c\theta_2) \end{matrix} &
\begin{matrix} c\eta_3\, s\varsigma_3 \\ +\varepsilon \begin{bmatrix} r_2 (c\theta_2\, c\varsigma_3 + s\theta_2\, s\eta_3\, s\varsigma_3) \\ + s_2 (s\theta_2\, c\varsigma_3 - c\theta_2\, s\eta_3\, s\varsigma_3) \end{bmatrix} \end{matrix} &
\begin{matrix} c\eta_3\, c\varsigma_3 \\ -\varepsilon \begin{bmatrix} r_2 (c\theta_2\, s\varsigma_3 - s\theta_2\, s\eta_3\, c\varsigma_3) \\ + s_2 (s\theta_2\, s\varsigma_3 + c\theta_2\, s\eta_3\, c\varsigma_3) \end{bmatrix} \end{matrix}
\end{bmatrix}
=
$$

$$
\begin{bmatrix}
\begin{matrix} c\theta_1 \\ -\varepsilon s_4\, c\alpha_4\, s\theta_1 \end{matrix} &
\begin{matrix} c\alpha_4\, s\theta_1 \\ -\varepsilon (s_4\, c\theta_1 - a_4\, s\alpha_4\, s\theta_1) \end{matrix} &
\begin{matrix} s\alpha_4\, s\theta_1 \\ +\varepsilon a_4\, c\alpha_4\, s\theta_1 \end{matrix} \\[2ex]
\begin{matrix} -c\alpha_1\, s\theta_1 \\ +\varepsilon s_4 (s\alpha_1\, s\alpha_4 - c\alpha_1\, c\alpha_4\, c\theta_1) \end{matrix} &
\begin{matrix} -s\alpha_1\, s\alpha_4 + c\alpha_1\, c\alpha_4\, c\theta_1 \\ +\varepsilon \begin{bmatrix} a_4 (s\alpha_1\, c\alpha_4 + c\alpha_1\, s\alpha_4\, c\theta_1) \\ - s_4\, c\alpha_1\, s\theta_1 \end{bmatrix} \end{matrix} &
\begin{matrix} s\alpha_1 c\alpha_4 + c\alpha_1\, s\alpha_4\, c\theta_1 \\ -\varepsilon a_4 (s\alpha_1\, s\alpha_4 - c\alpha_1\, c\alpha_4\, c\theta_1) \end{matrix} \\[2ex]
\begin{matrix} s\alpha_1\, s\theta_1 \\ +\varepsilon s_4 (c\alpha_1\, s\alpha_4 + s\alpha_1\, c\alpha_4\, c\theta_1) \end{matrix} &
\begin{matrix} -c\alpha_1\, s\alpha_4 - s\alpha_1\, c\alpha_4\, c\theta_1 \\ -\varepsilon \begin{bmatrix} a_4 (c\alpha_1\, c\alpha_4 - s\alpha_1\, s\alpha_4\, c\theta_1) \\ - s_4\, s\alpha_1\, s\theta_1 \end{bmatrix} \end{matrix} &
\begin{matrix} c\alpha_1\, c\alpha_4 - s\alpha_1\, s\alpha_4\, c\theta_1 \\ -\varepsilon a_4 (c\alpha_1\, s\alpha_4 + s\alpha_1\, c\alpha_4\, c\theta_1) \end{matrix}
\end{bmatrix}
\tag{265}
$$

Equating the primary components of the 2-1 elements of eq.(265)

$$s\theta_2\, c\eta_3 = -c\alpha_1\, s\theta_1 \tag{266}$$

and primary components of the 1-1 elements of eq.(265)

$$c\theta_2\, c\eta_3 = c\theta_1 \tag{267}$$

Combining eq.(266) and eq.(267),

$$\tan\theta_2 = \frac{-c\alpha_1\, s\theta_1}{c\theta_1} \tag{268}$$

From the primary component of 3-1 element of eq.(265)

$$s\eta_3 = -s\alpha_1\, s\theta_1 \tag{269}$$

Adding the product of $c\theta_2$ and the primary component of the 1-1 element of eq.(265) and the product of $s\theta_2$ and the primary component of the 2-1 element of eq.(265) results in

$$c\eta_3 = c\theta_1\, c\theta_2 - c\alpha_1\, s\theta_1\, s\theta_2 \tag{270}$$

Combination of eqs.(269) and (270) yields

$$\tan\eta_3 = \frac{-s\alpha_1\, s\theta_1}{c\theta_1\, c\theta_2 - c\alpha_1\, s\theta_1\, s\theta_2} \tag{271}$$

The result of taking the quotient of the primary components of the 3-2 and 3-3 elements of eq.(265) is

$$\tan\varsigma_3 = \frac{-c\alpha_1\, s\alpha_4 - s\alpha_1\, c\alpha_4\, c\theta_1}{c\alpha_1\, c\alpha_4 - s\alpha_1\, s\alpha_4\, c\theta_1} \tag{272}$$

Eqs.(268,271,272) can be used to determine the intermediate rotation displacements with consideration of the signs of the numerators and denominators indicating the proper quadrants and the angles.

The dual component of the 2-3 element of eq.(265) gives

$$r_2 = \frac{a_4\left(s\alpha_1\, s\alpha_4 - c\alpha_1\, c\alpha_4\, c\theta_1\right)}{c\eta_3\, c\varsigma_3} \tag{273}$$

From the dual component of element 1-3 of eq.(265),

$$s_2 = \frac{a_4\, c\alpha_4\, s\theta_1}{c\eta_3\, c\varsigma_3} \tag{274}$$

The output translational displacement can be obtained from consideration of the dual component of the 2-1 element of eq.(265).

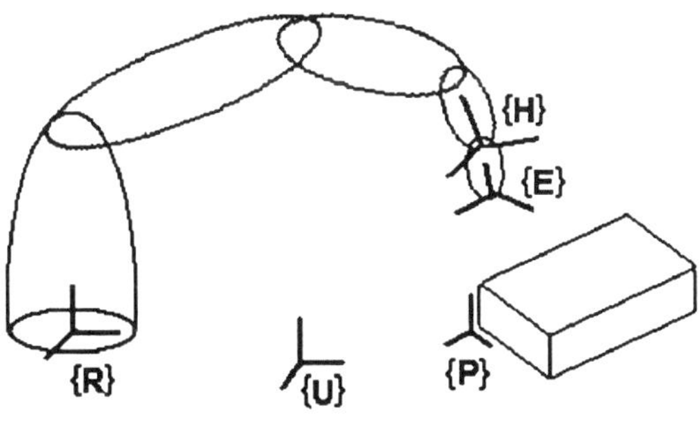

Figure 26 Universe, robot base, hand, end effector and workpiece frames.

Robotic Systems (Open-Loop Mechanisms)

Given a robot which is to perform a certain operation on a workpiece, it is desired to determine the displacements of the robot's joints.

In this robot system, the frames, which are shown in Figure 26, are

{U} = Universe (World)

{R} = Robot base

{H} = Hand

{E} = End effector (tool tip)

{P} = Part (workpiece)

For an open-loop mechanism such as this robot system, we can set up an equation analogous to the loop-closure equation by filling in the missing portions of the loop

$$ {}^{U}_{R}\hat{T}\ {}^{R}_{H}\hat{T}\ {}^{H}_{E}\hat{T}\ {}^{E}_{P}\hat{T}\ {}^{P}_{U}\hat{T} = \hat{I} \tag{275} $$

where

${}^{U}_{R}\hat{T}$ represents the situation of the robot within the factory.

${}^{H}_{E}\hat{T}$ represents the situation of the tool tip relative to the hand (depends on the tooling)

${}^{E}_{P}\hat{T}$ represents the situation of the tool tip relative to the workpiece (depends on the desired operation)

${}^{P}_{U}\hat{T}$ represents the situation of the workpiece within the factory.

${}^{R}_{H}\hat{T}$ represents the configuration of the robot, and must be obtained as a first step in determining the joint displacements.

Eq.(275) can be rearranged to yield the transform which must be determined:

$$ {}^{R}_{H}\hat{T} = {}^{R}_{U}\hat{T}\ {}^{U}_{P}\hat{T}\ {}^{P}_{E}\hat{T}\ {}^{E}_{H}\hat{T} \tag{276} $$

The right-hand side of eq.(276) consists of the robot link dimensions and the unknown joint displacements while the left-hand side of eq.(276) is a constant which can be evaluated by inspecting the situation of the hand.

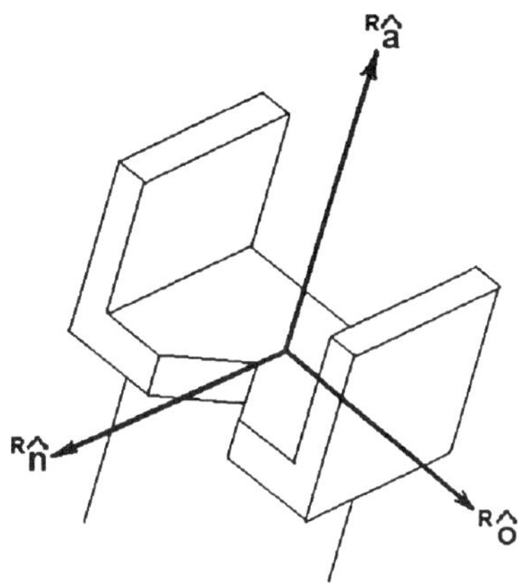

Figure 27 Vectors defining the hand frame.

Now, let us take a look at the matrix which specifies the hand frame.

$$^R_H\hat{T} = \begin{bmatrix} ^R\hat{a} & ^R\hat{o} & ^R\hat{n} \end{bmatrix} \tag{277}$$

The component unit vectors, shown in Figure 27, are

$$^R\hat{a} = \begin{bmatrix} ^R\hat{a}_x \\ ^R\hat{a}_y \\ ^R\hat{a}_z \end{bmatrix} = \text{approach vector} \tag{278}$$

$$^R\hat{o} = \begin{bmatrix} ^R\hat{o}_x \\ ^R\hat{o}_y \\ ^R\hat{o}_z \end{bmatrix} = \text{orientation vector} \tag{279}$$

$$^R\hat{n} = \begin{bmatrix} ^R\hat{n}_x \\ ^R\hat{n}_y \\ ^R\hat{n}_z \end{bmatrix} = \text{normal vector} \tag{280}$$

Because these vectors comprise a coordinate frame, these vectors are related to each other such that

$$^R\hat{n} \times {}^R\hat{o} = {}^R\hat{a} \tag{281}$$

$$^R\hat{a} \times {}^R\hat{n} = {}^R\hat{o} \tag{282}$$

$$^R\hat{o} \times {}^R\hat{a} = {}^R\hat{n} \tag{283}$$

Using the concepts of the approach, orientation and normal vectors, the matrix ${}^{R}_{H}\hat{T}$ can be easily constructed and can be used in eq.(276) where the other matrices contain the dimensions and joint displacements of the robot.

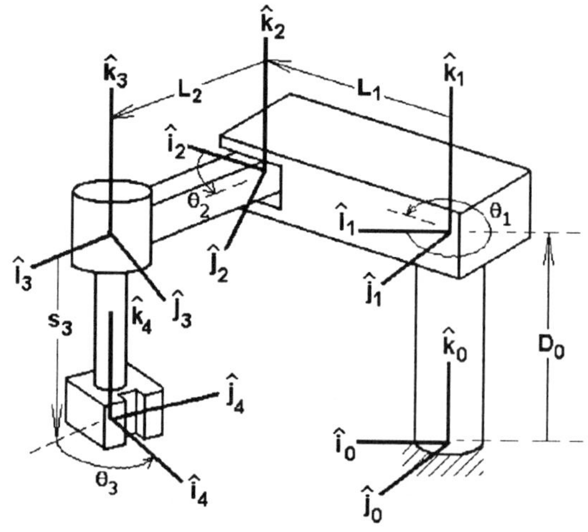

Figure 28 Robot used in example. Notice that the value of displacement s_3 will be negative for the configuration shown. The ground is link 0.

Example

The displacements of the robot shown in Figure 28 are to be analyzed. Consider frames {3} and {4}. Axes $\hat{k}_3$ and $\hat{k}_4$ are collinear, so link length $a_4 = 0$. The angle between axes $\hat{k}_3$ and $\hat{k}_4$ is link twist $\alpha_4 = 0$. The joint offset at joint 3 is length s_3 (the direction of displacement shown is the negative direction). The joint rotation at joint 3 is angle θ_3.

The Denavit-Hartenberg parameters for this robot are given in Table 7.

Table 7 - Hartenberg-Denavit Parameters for Robot Shown in Figure 28				
i	θ_i	s_i	α_i	a_i
0	0	D_0	0	0
1	θ_1	0	0	L_1
2	θ_2	0	0	L_2
3	θ_3	s_3	0	0

These values can be plugged into our standard link-joint transformation

$$ {}^{0}_{1}\hat{M}\,{}^{1}_{2}\hat{M}\,{}^{2}_{3}\hat{M}\,{}^{3}_{4}\hat{M} = {}^{R}_{H}\hat{T} = {}^{0}_{4}\hat{M} $$

$$ = $$

$$ \begin{bmatrix} 1 & \varepsilon D_0 & 0 \\ -\varepsilon D_0 & 1 & 0 \\ 0 & 0 & 1 \end{bmatrix} \begin{bmatrix} c\theta_1 & -s\theta_1 & \varepsilon L_1\, s\theta_1 \\ s\theta_1 & c\theta_1 & -\varepsilon L_1\, c\theta_1 \\ 0 & \varepsilon L_1 & 1 \end{bmatrix} $$

$$ \times $$

$$ \begin{bmatrix} c\theta_2 & -s\theta_2 & \varepsilon L_2\, s\theta_2 \\ s\theta_2 & c\theta_2 & -\varepsilon L_2\, c\theta_2 \\ 0 & \varepsilon L_2 & 1 \end{bmatrix} \begin{bmatrix} c\theta_3 - \varepsilon s_3\, s\theta_3 & -s\theta_3 - \varepsilon s_3\, c\theta_3 & 0 \\ s\theta_3 + \varepsilon s_3\, c\theta_3 & c\theta_3 - \varepsilon s_3\, s\theta_3 & 0 \\ 0 & 0 & 1 \end{bmatrix} $$

$$ = $$

$$ \begin{bmatrix} c\theta_{123} + \varepsilon(D_0 - s_3)s\theta_{123} & -s\theta_{123} + \varepsilon(D_0 - s_3)c\theta_{123} & \varepsilon(L_1\, s\theta_1 + L_2\, s\theta_{12}) \\ s\theta_{123} - \varepsilon(D_0 - s_3)c\theta_{123} & c\theta_{123} + \varepsilon(D_0 - s_3)s\theta_{123} & -\varepsilon(L_1\, c\theta_1 + L_2\, c\theta_{12}) \\ \varepsilon(L_2\, s\theta_3 + L_1\, s\theta_{23}) & \varepsilon(L_2\, c\theta_3 + L_1\, c\theta_{23}) & 1 \end{bmatrix} $$

where $\theta_{123} = \theta_1 + \theta_2 + \theta_3$, $\theta_{12} = \theta_1 + \theta_2$, and $\theta_{23} = \theta_2 + \theta_3$.

Plugging in the joint displacements to obtain the situation of the hand frame is called the "forward transformation" and is a simple task.

Obtaining the joint displacements when given a matrix with numeric elements is called the "inverse transformation." It is often a mathematically difficult task.

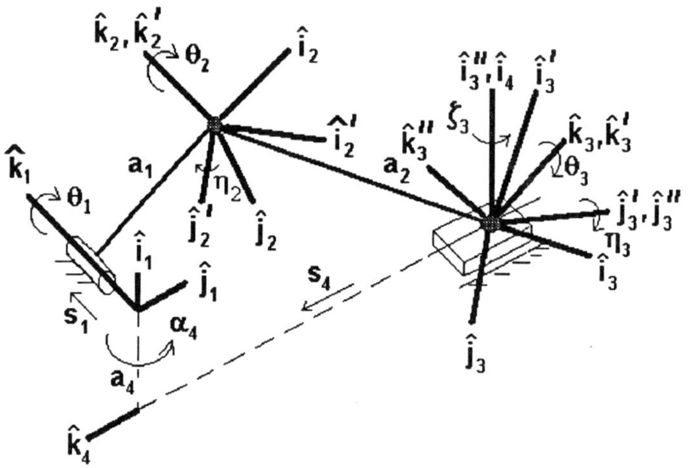

Figure 29 Displacements of the generalized slider-crank mechanism.

Generalized Slider-Crank Mechanism

The generalized slider-crank mechanism is of the CSSP type as illustrated in Figure 29. The connection between the ground link and the input crank is cylindrical, allowing both sliding and rotation, two degrees of freedom. The coupler link is connected to the input crank and the slider by ball joints allowing rotation about all three axes. Only two rotations will be considered in the joint between the crank and the coupler since rotation about the axis through both ball joints of the coupler does no useful work and is said to represent a passive degree of freedom. The slider is connected to the ground by a prismatic joint.

This mechanism can be specialized to a planar slider-crank, RRRP mechanism by setting the sliding at the input shaft, s_1, to zero, considering the coupler joints as revolute, so that rotations η_2 and η_3 are set to zero, the rotation ζ_3 fixed so that it is considered as a link twist $\alpha_3 = {}^{\pi}/_{2}$ and the ground-link twist α_4 is set to a value of ${}^{3\pi}/_{2}$.

For this mechanism, the condition of loop closure is

$$ {}^{1}_{2}\hat{M}\,{}^{2}_{3}\hat{L}\,{}^{3}_{4}\hat{L}\,{}^{4}_{1}\hat{M} = \hat{I} \tag{284}$$

Table 8 - Generalized Slider-Crank Mechanism

i	θ_i	s_i	η_i	ζ	α_i	a_i
1	θ_1	s_1	0	0	0	a_1
2	θ_2	0	η_2	0	0	a_2
3	θ_3	0	η_3	ζ_3	0	0
4	0	s_4	0	0	α_4	a_4

where ${}^{i}_{i+1}\hat{\mathrm{M}}$ is the transform for a link with a cylindrical joint on the proximal end and where ${}^{i}_{i+1}\hat{\mathrm{L}}$ is the transform for a link with a ball joint on the proximal end.

Considering the Denavit-Hartenberg parameters, as given in Table 8, the individual transform matrices are

$$
{}^{1}_{2}\hat{\mathrm{M}} = \begin{bmatrix} c\theta_1 & -s\theta_1 & 0 \\ s\theta_1 & c\theta_1 & 0 \\ 0 & 0 & 1 \end{bmatrix} + \varepsilon \begin{bmatrix} -s_1\,s\theta_1 & -s_1\,c\theta_1 & a_1\,s\theta_1 \\ s_1\,c\theta_1 & -s_1\,s\theta_1 & -a_1\,c\theta_1 \\ 0 & a_1 & 0 \end{bmatrix} \tag{285}
$$

$$
{}^{2}_{3}\hat{\mathrm{L}} = \begin{bmatrix} c\theta_2\,c\eta_2 & -s\theta_2 & c\theta_2\,s\eta_2 \\ s\theta_2\,c\eta_2 & c\theta_2 & s\theta_2\,s\eta_2 \\ -s\eta_2 & 0 & c\eta_2 \end{bmatrix} + \varepsilon \begin{bmatrix} 0 & a_2\,c\theta_2\,s\eta_2 & a_2\,s\theta_2 \\ 0 & a_2\,s\theta_2\,s\eta_2 & -a_2\,c\theta_2 \\ 0 & a_2\,c\eta_2 & 0 \end{bmatrix} \tag{286}
$$

$$
{}^{3}_{4}\hat{\mathrm{L}} = \begin{bmatrix} c\theta_3\,c\eta_3 & -s\theta_3\,c\zeta_3 + c\theta_3\,s\eta_3\,s\zeta_3 & s\theta_3\,s\zeta_3 + c\theta_3\,s\eta_3\,c\zeta_3 \\ s\theta_3\,c\eta_3 & c\theta_3\,c\zeta_3 + s\theta_3\,s\eta_3\,s\zeta_3 & -c\theta_3\,s\zeta_3 + s\theta_3\,s\eta_3\,c\zeta_3 \\ -s\eta_3 & c\eta_3\,s\zeta_3 & c\eta_3\,c\zeta_3 \end{bmatrix} \tag{287}
$$

$$
{}^{4}_{1}\hat{\mathrm{M}} = \begin{bmatrix} 1 & 0 & 0 \\ 0 & c\alpha_4 & -s\alpha_4 \\ 0 & s\alpha_4 & c\alpha_4 \end{bmatrix} + \varepsilon \begin{bmatrix} 0 & -s_4\,c\alpha_4 & s_4\,s\alpha_4 \\ s_4 & -a_4\,s\alpha_4 & -a_4\,c\alpha_4 \\ 0 & a_4\,c\alpha_4 & -a_4\,s\alpha_4 \end{bmatrix} \tag{288}
$$

Uicker-Denavit-Hartenberg Solution Scheme

Eq.(284) can be solved for the displacements by a Newton-Raphson iteration scheme developed by Uicker, Denavit and Hartenberg (1964). The UDH method, originally designed for homogeneous coordinate-transformations, was adopted to dual-number methods by Fischer and Paul (1987).

A first-order expansion of eq.(284) is

$$\left[{}^{1}_{2}\hat{M}\right]\left[{}^{2}_{3}\hat{L} + \hat{Q}_{L\theta}\,{}^{2}_{3}\hat{L}\delta\theta_2 + \hat{Q}_{L\eta}\,{}^{2}_{3}\hat{L}\delta\eta_2\right]\left[{}^{3}_{4}\hat{L} + \hat{Q}_{L\theta}\,{}^{3}_{4}\hat{L}\delta\theta_3 + \hat{Q}_{L\eta}\,{}^{3}_{4}\hat{L}\delta\eta_3 + \hat{Q}_{L\zeta}\,{}^{3}_{4}\hat{L}\delta\zeta_3\right]\left[{}^{4}_{1}\hat{M} + \hat{Q}_{Ms}\,{}^{4}_{1}\hat{M}\delta s_4\right] = \hat{I} \quad (289)$$

Matrices ${}^{1}_{2}\hat{M}$, ${}^{2}_{3}\hat{L}$, ${}^{3}_{4}\hat{L}$, ${}^{4}_{1}\hat{M}$ are each evaluated at

θ_2, η_2, θ_3, η_3, s_4 which are the *guesses* of the displacement variables,

θ_1 which is the *input* displacement,

$\delta\theta_2$, $\delta\eta_2$, $\delta\theta_3$, $\delta\eta_3$, $\delta\zeta_3$, δs_4 which are the *errors* between the guesses and the exact values of the displacements

and

$\hat{Q}_{L\theta}$, $\hat{Q}_{L\eta}$, $\hat{Q}_{L\zeta}$, $\hat{Q}_{Ms}$ are partial-derivative operators corresponding to the link-joint transforms such that

$$\hat{Q}_{M\theta} = \begin{bmatrix} 0 & -1 & 0 \\ 1 & 0 & 0 \\ 0 & 0 & 0 \end{bmatrix} \quad (290)$$

$$\hat{Q}_{Ms} = \begin{bmatrix} 0 & -\varepsilon & 0 \\ \varepsilon & 0 & 0 \\ 0 & 0 & 0 \end{bmatrix} \quad (291)$$

$$\hat{Q}_{L\theta} = \begin{bmatrix} 0 & -1 & 0 \\ 1 & 0 & 0 \\ 0 & 0 & 0 \end{bmatrix} \quad (292)$$

$$\hat{Q}_{L\eta} = \begin{bmatrix} 0 & 0 & c\hat{\theta} \\ 0 & 0 & s\hat{\theta} \\ -c\hat{\theta} & -s\hat{\theta} & 0 \end{bmatrix} \quad (293)$$

$$\hat{Q}_{L\zeta} = \begin{bmatrix} 0 & s\hat{\eta} & s\hat{\theta}c\hat{\eta} \\ -s\hat{\eta} & 0 & -c\hat{\theta}c\hat{\eta} \\ -s\hat{\theta}c\hat{\eta} & -c\hat{\theta}c\hat{\eta} & 0 \end{bmatrix} \quad (294)$$

Eq.(289) can be expanded and put in a form which is computationally advantageous

$$\hat{H}_2^{M\theta}{}_2\delta\theta_2 + \hat{H}_2^{L\eta}\delta\eta_2 + \hat{H}_3^{L\theta}\delta\theta_3 + \hat{H}_3^{L\eta}\delta\eta_3 + \hat{H}_3^{L\zeta}\delta\zeta_3 + \hat{H}_4^{Ms}\delta s_4 = \hat{B}^T - \hat{I} \quad (295)$$

where

$$\hat{H}_2^{L\theta} = {}_2^1\hat{M}\hat{Q}_{L\theta}\,{}_2^1\hat{M}^T \tag{296}$$

$$\hat{H}_2^{L\eta} = {}_2^1\hat{M}\hat{Q}_{L\eta}\,{}_2^1\hat{M}^T \tag{297}$$

$$\hat{H}_3^{L\theta} = \left({}_2^1\hat{M}_3^2\hat{L}\right)\hat{Q}_{L\theta}\left({}_2^1\hat{M}_3^2\hat{L}\right)^T \tag{298}$$

$$\hat{H}_3^{L\eta} = \left({}_2^1\hat{M}_3^2\hat{L}\right)\hat{Q}_{L\eta}\left({}_2^1\hat{M}_3^2\hat{L}\right)^T \tag{299}$$

$$\hat{H}_3^{L\zeta} = \left({}_2^1\hat{M}_3^2\hat{L}\right)\hat{Q}_{L\zeta}\left({}_2^1\hat{M}_3^2\hat{L}\right)^T \tag{300}$$

$$\hat{H}_4^{Ms} = \left({}_2^1\hat{M}_3^2\hat{L}_4^3\hat{L}\right)\hat{Q}_{Ms}\left({}_2^1\hat{M}_3^2\hat{L}_4^3\hat{L}\right)^T \tag{301}$$

In general,

$$\hat{H}_i^x = \left({}_2^1\hat{T}_3^2\hat{T}\ldots{}_i^{i+1}\hat{T}\right)\hat{Q}_x\left({}_2^1\hat{T}_3^2\hat{T}\ldots{}_i^{i+1}\hat{T}\right)^T \tag{302}$$

where ${}_i^{i+1}\hat{T}$ is the appropriate joint-link transform and $\hat{Q}_x$ is the appropriate partial-derivative operator.

Also in eq.(295),

$$\hat{B} = {}_2^1\hat{M}\;{}_3^2\hat{L}\;{}_4^3\hat{L}\;{}_1^4\hat{M} \tag{303}$$

In general,

$$\hat{B} = {}_2^1\hat{T}\;{}_3^2\hat{T}\ldots{}_n^{n-1}\hat{T} \tag{304}$$

where matrix ${}_{i+1}^{\;i}\hat{T}$ is the appropriate joint-link transform for link i of the n links in a closed-loop mechanism.

Eq.(295) can be condensed to the compact form

$$\hat{F} = \hat{B}^T - \hat{I} \tag{305}$$

Writing out each term of eq.(305) produces

$$\begin{bmatrix} \hat{F}_{11} & \hat{F}_{12} & \hat{F}_{13} \\ \hat{F}_{21} & \hat{F}_{22} & \hat{F}_{23} \\ \hat{F}_{31} & \hat{F}_{32} & \hat{F}_{33} \end{bmatrix} = \begin{bmatrix} \hat{B}_{11}-1 & \hat{B}_{21} & \hat{B}_{31} \\ \hat{B}_{12} & \hat{B}_{22}-1 & \hat{B}_{32} \\ \hat{B}_{13} & \hat{B}_{23} & \hat{B}_{33}-1 \end{bmatrix} \tag{306}$$

This matrix equation comprises nine dual equations of which, because of the orthogonality of a dual-number coordinate-transformation matrix, only three are independent.

Using the three terms in the upper right-hand triangle of each matrix, and separating them into a set of primary and a set of dual equations produces

$$\begin{aligned} F_{p12} &= B_{p21} \\ F_{p13} &= B_{p31} \\ F_{p23} &= B_{p32} \end{aligned} \tag{307}$$

and

$$\begin{aligned} F_{d12} &= B_{d21} \\ F_{d13} &= B_{d31} \\ F_{d23} &= B_{d32} \end{aligned} \tag{308}$$

where subscripts p and d denote primary and dual terms.

The primary set, eq.(307), however, does not include the constraint that the diagonal terms must each be equal to unity at closure. This constraint can be included as three additional equations as described by Uicker, Denavit and Hartenberg (1964) or through modification of the current equation set as described in Sandor and Erdman (1984).

Adopting the second method modifies eq.(307) to

$$\begin{aligned} F_{p12} &= B_{p12} + B_{p11} + B_{p22} - 2 \\ F_{p13} &= B_{p31} + B_{p11} + B_{p33} - 2 \\ F_{p23} &= B_{p32} + B_{p22} + B_{p33} - 2 \end{aligned} \tag{309}$$

The left-hand side of eqs.(308) and (309) contain the linear combination of the respective elements of the $\hat{\mathrm{H}}$ matrices and the error terms as can be seen in the left-hand side of eq.(295).

Eqs.(308) and (309) can be combined as partitions of a single matrix of primary-number equations having the form

$$\left[\frac{\mathrm{A}_p}{\mathrm{A}_d}\right]\{\Delta\} = \left\{\frac{\mathrm{V}_p}{\mathrm{V}_d}\right\} \tag{310}$$

where

$$\mathrm{A} = \begin{bmatrix} \mathrm{A}_p \\ \hdashline \mathrm{A}_d \end{bmatrix} = \begin{bmatrix} H_{2p12}^{L\theta} & H_{2p12}^{L\eta} & H_{3p12}^{L\theta} & H_{3p12}^{L\eta} & H_{3p12}^{L\zeta} & H_{3p12}^{Ms} \\ H_{2p13}^{L\theta} & H_{2p13}^{L\eta} & H_{3p13}^{L\theta} & H_{3p12}^{L\eta} & H_{3p13}^{L\zeta} & H_{3p13}^{Ms} \\ H_{2p23}^{L\theta} & H_{2p23}^{L\eta} & H_{3p23}^{L\theta} & H_{3p12}^{L\eta} & H_{3p23}^{L\zeta} & H_{3p23}^{Ms} \\ \hdashline H_{2d12}^{L\theta} & H_{2d12}^{L\eta} & H_{3d12}^{L\theta} & H_{3d12}^{L\eta} & H_{3d12}^{L\zeta} & H_{3d12}^{Ms} \\ H_{2d13}^{L\theta} & H_{2d13}^{L\eta} & H_{3d13}^{L\theta} & H_{3d13}^{L\eta} & H_{3d13}^{L\zeta} & H_{3d13}^{Ms} \\ H_{2d23}^{L\theta} & H_{2d23}^{L\eta} & H_{3d23}^{L\theta} & H_{3d23}^{L\eta} & H_{3d23}^{L\zeta} & H_{3d23}^{Ms} \end{bmatrix} \tag{311}$$

$$\mathrm{V} = \begin{Bmatrix} \mathrm{V}_p \\ \hdashline \mathrm{V}_d \end{Bmatrix} = \begin{Bmatrix} B_{p21} + B_{p11} + B_{p22} - 2 \\ B_{p31} + B_{p11} + B_{p33} - 2 \\ B_{p32} + B_{p22} + B_{p33} - 2 \\ \hdashline B_{d21} \\ B_{d31} \\ B_{d32} \end{Bmatrix} \tag{312}$$

and

$$\Delta = \begin{Bmatrix} \delta\theta_2 \\ \delta\eta_2 \\ \delta\theta_3 \\ \delta\eta_3 \\ \delta\zeta_3 \\ \delta s_4 \end{Bmatrix} \tag{313}$$

Eq.(310), written as

$$\mathrm{A}\Delta = \mathrm{V} \tag{314}$$

is solved for the error Δ.

In many mechanisms which will be considered, the coefficient matrix in eq.(314) is non-square. Even so, the equation set will have solutions if the rank of the coefficient matrix equals the number of its columns. A solution can be obtained by manipulating eq.(314) in the following manner:

$$\mathrm{A}^{\mathrm{T}}\mathrm{A}\Delta = \mathrm{A}^{\mathrm{T}}\mathrm{V} \tag{315}$$

$$\left(\mathrm{A}^{\mathrm{T}}\mathrm{A}\right)^{-1}\mathrm{A}^{\mathrm{T}}\mathrm{A}\Delta = \left(\mathrm{A}^{\mathrm{T}}\mathrm{A}\right)^{-1}\mathrm{A}^{\mathrm{T}}\mathrm{V} \tag{316}$$

$$\Delta = \left(\mathrm{A}^{\mathrm{T}}\mathrm{A}\right)^{-1}\mathrm{A}^{\mathrm{T}}\mathrm{V} \tag{317}$$

The guesses of the displacement variables are revised by the error terms Δ. The new set of joint displacements is used in the next iteration to produce a new set of error terms. The process is repeated until convergence has occurred, when the error terms are smaller than some pre-selected tolerance value.

After obtaining the output and intermediate displacements for a given set of input displacements, the input variables are set to new values and the entire process repeated. Because the Newton-Raphson method is subject to failure if the initial guess is poor, the step in input values must be reasonably small if the results of the prior calculations are to be used as the initial guess.

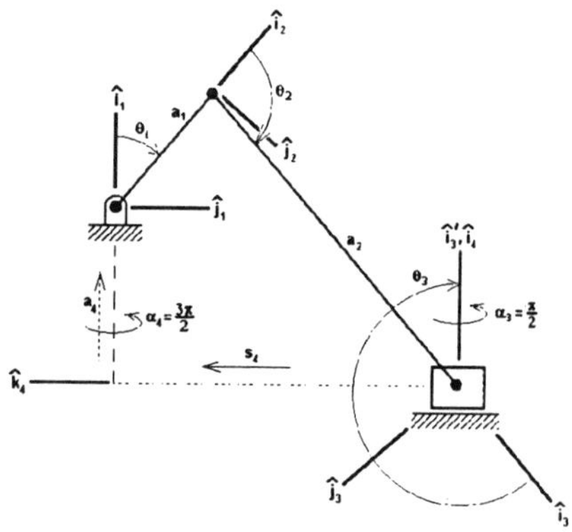

Figure 30 Displacements of the planar slider-crank mechanism.

Planar Slider-Crank Mechanism

Consider an RRRP mechanism where the links are constrained to lie within a single plane in space. This is the common planar slider-crank mechanism where the ground link has an offset of length a_4, the input crank has length a_1 and the coupler has length a_2 as shown in Figure 30.

Table 9 - Planar Slider-Crank Mechanism

i	θ_i	s_i	α_i	a_i
1	θ_1	0	0	a_1
2	θ_2	0	0	a_2
3	θ_3	0	$\pi/2$	0
4	0	s_4	$3\pi/2$	a_4

The condition of loop closure requires that

$$ {}_{2}^{1}\hat{\mathrm{M}}\ {}_{3}^{2}\hat{\mathrm{M}}\ {}_{4}^{3}\hat{\mathrm{M}}\ {}_{1}^{4}\hat{\mathrm{M}} = \hat{\mathrm{I}} \tag{318} $$

The values of the Denavit-Hartenberg parameters in Table 9 substituted into the joint-link transforms yields

$$ {}_{2}^{1}\hat{\mathrm{M}} = \begin{bmatrix} \mathrm{c}\theta_1 & -\mathrm{s}\theta_1 & \varepsilon a_1\,\mathrm{s}\theta_1 \\ \mathrm{s}\theta_1 & \mathrm{c}\theta_1 & -\varepsilon a_1\,\mathrm{c}\theta_1 \\ 0 & \varepsilon a_1 & 1 \end{bmatrix} \tag{319} $$

$$ {}_{3}^{2}\hat{\mathrm{M}} = \begin{bmatrix} \mathrm{c}\theta_2 & -\mathrm{s}\theta_2 & \varepsilon a_2\,\mathrm{s}\theta_2 \\ \mathrm{s}\theta_2 & \mathrm{c}\theta_2 & -\varepsilon a_2\,\mathrm{c}\theta_2 \\ 0 & \varepsilon a_2 & 1 \end{bmatrix} \tag{320} $$

$$ {}_{4}^{3}\hat{\mathrm{M}} = \begin{bmatrix} \mathrm{c}\theta_3 & 0 & \mathrm{s}\theta_3 \\ \mathrm{s}\theta_3 & 0 & -\mathrm{c}\theta_3 \\ 0 & 1 & 0 \end{bmatrix} \tag{321} $$

$$ {}_{1}^{4}\hat{\mathrm{M}} = \begin{bmatrix} 1 & 0 & -\varepsilon s_4 \\ \varepsilon s_4 & \varepsilon a_4 & 1 \\ 0 & -1 & \varepsilon a_4 \end{bmatrix} \tag{322} $$

Eq.(318) is adjusted to the form

$$ {}_{3}^{2}\hat{\mathrm{M}}\ {}_{4}^{3}\hat{\mathrm{M}} = {}_{2}^{1}\hat{\mathrm{M}}^{\mathrm{T}}\ {}_{1}^{4}\hat{\mathrm{M}}^{\mathrm{T}} \tag{323} $$

and the joint-link transforms are substituted to obtain

$$ \begin{bmatrix} \mathrm{c}\theta_{23} & \varepsilon a_2\,\mathrm{s}\theta_2 & \mathrm{s}\theta_{23} \\ \mathrm{s}\theta_{23} & -\varepsilon a_2\,\mathrm{c}\theta_2 & -\mathrm{c}\theta_{23} \\ \varepsilon a_2\,\mathrm{s}\theta_3 & 1 & -\varepsilon a_2\,\mathrm{c}\theta_3 \end{bmatrix} = \begin{bmatrix} \mathrm{c}\theta_1 & \varepsilon(a_4\,\mathrm{s}\theta_1 + s_4\,\mathrm{c}\theta_1) & -\mathrm{s}\theta_1 \\ -\mathrm{s}\theta_1 & \varepsilon(a_1 + a_4\,\mathrm{c}\theta_1 - s_4\,\mathrm{s}\theta_1) & -\mathrm{c}\theta_1 \\ \varepsilon(a_1\,\mathrm{s}\theta_1 - s_4) & 1 & \varepsilon(a_1\,\mathrm{c}\theta_1 + a_4) \end{bmatrix} \tag{324} $$

Equating the 3-1 and 3-3 elements,

$$ a_2\,\mathrm{s}\theta_3 = a_1\,\mathrm{s}\theta_1 - s_4 \tag{325} $$

$$ -a_2\,\mathrm{c}\theta_3 = a_1\,\mathrm{c}\theta_1 + a_4 \tag{326} $$

Squaring each side of these equations and adding together yields

$$a_2^2 = a_1^2 - 2a_1 s_4 \,\mathrm{s}\theta_1 + 2a_1 a_4 \,\mathrm{c}\theta_1 + s_4^2 + a_4^2 \tag{327}$$

This is manipulated to solve for the slider translation in terms of the input rotation and link lengths:

$$s_4 = a_1 \,\mathrm{s}\theta_1 \pm \sqrt{a_1^2 \,\mathrm{s}^2\theta_1 - a_1^2 + a_2^2 - a_4^2 - 2a_1 a_4 \,\mathrm{c}\theta_1} \tag{328}$$

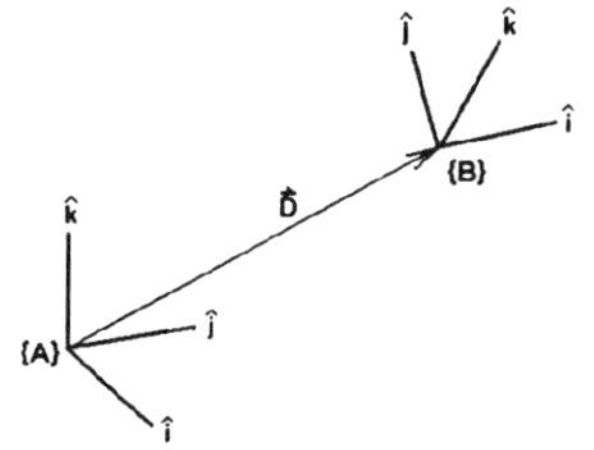

Figure 31 The origin-displacement equation produces the vector $\vec{D}$ between the origins of frames {A} and {B} given the dual-number coordinate-transformation matrix relating the two frames.

Origin-Displacement Equation

The "origin-displacement equation" referred to in this discussion is a relationship in dual-number mathematics which allows the relative position of two coordinate frames to be recovered from the dual-number coordinate-transformation matrix between them. It was developed by Hsia and Yang (1981) who referred to it as the "principle of transference" and applied it to the determination of instantaneous invariants. Later it was applied to robotic-manipulator kinematics by Pennock and Yang (1985).

In this text, this relationship will be referred to as the "origin-displacement equation" rather than "principle of transference" to avoid confusion with the principle of transference, sometimes referred to as the transfer principle, formulated by Study and Kotelnikov and developed by Dimentberg which notes that trigonometric relationships between primary angles also hold for dual angles.

Consider the two frames {A} and {B} as are shown in Figure 31. Let

[R] = 3×3 primary-number matrix describing change in orientation of frame {B} relative to frame {A}

$\vec{D}$ = 3×1 primary-number matrix describing displacement of origin of frame {B} relative to frame {A}

Now, using our definition of a dual vector,

$$^{A}\hat{x}_B = \vec{x}_B + \varepsilon\left(\vec{D} \times \vec{x}_B\right) \tag{329}$$

Now, the vector representing vector $\vec{x}_A$ rotated into alignment with frame {B} is

$$\vec{x}_B = [R]\vec{x}_A \tag{330}$$

and

$$\vec{D} \times \vec{x}_B = [D]\vec{x}_B \tag{331}$$

where

$$[D] = \begin{bmatrix} 0 & -d_3 & d_2 \\ d_3 & 0 & -d_1 \\ -d_2 & d_1 & 0 \end{bmatrix} \tag{332}$$

is the cross-product matrix of the vector

$$\vec{D} = \begin{Bmatrix} d_1 \\ d_2 \\ d_3 \end{Bmatrix} \tag{333}$$

So,

$$^{A}\vec{x}_B = [R]\vec{x}_A + \varepsilon[D][R]\vec{x}_A \tag{334}$$

Similar expressions can be written for axes $^{A}\vec{y}_B$ and $^{A}\vec{z}_B$, and taken together yield

$$\{B\} = [R]\{A\} + \varepsilon[D][R]\{A\} \tag{335}$$

$$= \left([B] + \varepsilon[D]\right)[R]\{A\} \tag{336}$$

$$= {}^{B}_{A}\hat{T}\{A\} \tag{337}$$

Thus,

$$^{B}_{A}\vec{T} = \left([I] + \varepsilon[D]\right)[R] \tag{338}$$

where

[I] = 3×3 identity matrix

[D] = 3×3 cross-product matrix of displacement vector $\vec{D}$

[R] = 3×3 rotation matrix

Now, matrix ${}^{B}_{A}\hat{T}$ can be decomposed into primary and dual components

$$ {}^{B}_{A}\hat{T} = U + \varepsilon V \tag{339}$$

Equating the primary components of eqs.(338) and (339), we obtain

$$ U = [R] \tag{340}$$

and equating the dual components we obtain

$$ V = [D][R] \tag{341}$$

Substituting eq.(340) into eq.(341) and manipulating, we obtain

$$ [D] = VU^{T} \tag{342}$$

This is the origin-displacement equation as formulated by Hsia and Yang (1981).

Given transform T=U+εV between two frames, we can recover in 3×3 cross-product matrix form the displacement vector between the origins given the 3×3 matrices U and V.

Application to Calculation of Translational Displacements

The origin-displacement equation can be used in an iterative process to calculate the translational displacements in a spatial mechanism if the rotational displacements have been already determined. This procedure is not based on formulating an equation which is then solved by some mathematical exercise but mimics the way a person would determine the translational displacements were he or she given a set of links to assemble.

The links of a spatial mechanism can be symbolized by 3×3 dual-number matrices

$$ {}^{i}_{i+1}\hat{T} = {}^{i}_{i+1}U + \varepsilon\, {}^{i}_{i+1}V \tag{343}$$

For a four-link closed-loop mechanism there are four joint-link transforms related by

$$ {}^{j}_{k}\hat{T}\, {}^{k}_{l}\hat{T}\, {}^{l}_{m}\hat{T}\, {}^{m}_{j}\hat{T} = B_m + \varepsilon C_m \tag{344}$$

where *jklm* is any of the sequences 1234, 2341, 3412 or 4123 and where the condition of loop closure requires that the primary component of the right-hand side is the identity matrix

$$\mathrm{B}_m = \mathrm{I} \tag{345}$$

and the dual component of the left-hand side is the null matrix

$$\mathrm{C}_m = [0] \tag{346}$$

Substitution of eq.(339) into eq.(344) yields

$$_k^j\mathrm{U}\ _l^k\mathrm{U}\ _m^l\mathrm{U}\ _j^m\mathrm{U} = \mathrm{B}_m \tag{347}$$

and

$$_k^j\mathrm{V}\ _l^k\mathrm{U}\ _m^l\mathrm{U}\ _j^m\mathrm{U} + {_k^j\mathrm{U}}\ _l^k\mathrm{V}\ _m^l\mathrm{U}\ _j^m\mathrm{U} + {_k^j\mathrm{U}}\ _l^k\mathrm{U}\ _m^l\mathrm{V}\ _j^m\mathrm{U} + {_k^j\mathrm{U}}\ _l^k\mathrm{U}\ _m^l\mathrm{U}\ _j^m\mathrm{V} = \mathrm{B}_m \tag{348}$$

If the proper values of the displacement are substituted into eq.(347) then

$$_k^j\mathrm{U}\ _l^k\mathrm{U}\ _m^l\mathrm{U}\ _j^m\mathrm{U} = \mathrm{I} \tag{349}$$

Eq.(349) contains only variables representing the rotational displacements.

Consider the situation in which the rotational displacements have already been determined, but the translational displacements, unless made accurately, when substituted into eq.(348) would result in $\mathrm{C}_m \neq [0]$.

The origin-displacement equation states that

$$\mathrm{D}_m = \mathrm{C}_m \mathrm{B}_m^{\mathrm{T}} \tag{350}$$

where

$$\mathrm{D}_m = \begin{bmatrix} 0 & -M_{m3} & M_{m2} \\ M_{m3} & 0 & -M_{m1} \\ -M_{m2} & M_{m1} & 0 \end{bmatrix} \tag{351}$$

The elements of matrix D_m are such that

$$\vec{\mathrm{M}}_m = \begin{Bmatrix} M_{m1} \\ M_{m2} \\ M_{m3} \end{Bmatrix} \tag{352}$$

is a vector locating the origin of frame {m-1} relative to frame {m} in terms of the unit vectors of frame {m-1}. Essentially this vector represents the “miss” in closing the mechanism at joint *m*. Because the orientations of the links are known, eq.(347) becomes eq.(349) and B_m=I can be substituted into eq.(350) to yield

$$D_m = C_m \tag{353}$$

Eq.(353) can be the basis for a scheme for determining the unknown translational displacements s_i in a mechanism where the rotational displacements θ_i are already known. The basic steps of the solution method are as follows.

First, an initial guess is made of the unknown translational displacements s_i, usually s_i=0.

Second, assembly of the mechanism is considered with the links unconnected at one joint on account of the errors in the translational displacements. The miss in closure $\vec{M}_m$ is calculated for each of the joints, using eq.(348) and then eq.(353), followed by eq.(351) and eq.(352). For a four-link closed-chain mechanism, the miss in closure $\vec{M}_1$ corresponding to the assembly where the chain is left open at joint 1 is shown in Figure 32, the miss in closure $\vec{M}_2$ corresponding to the assembly with the chain open at joint 2 is illustrated in Figure 33, the miss in closure $\vec{M}_3$ in the assembly that has the chain unconnected at joint 3 is shown in Figure 34, and the miss in closure $\vec{M}_4$ of the assembly with the chain open at joint 4 is depicted in Figure 35.

Third, eq.(350) expresses the miss in three coordinate directions, two perpendicular to the translational joint axis and one aligned with the translational joint axis. The amount of miss $\vec{M}_m$ can be minimized by sliding the link through a distance d_m such that

$$d_m = - M_{m3} \tag{354}$$

This is done for each joint in the mechanism and the miss $\vec{M}_m$ is recomputed. It will not vanish because nothing was done to eliminate the M_{m1} and M_{m2} components, but the total magnitude of the miss vector $\vec{M}_m$ will diminish. The process is reiterated until closure converges to within a desired accuracy value.

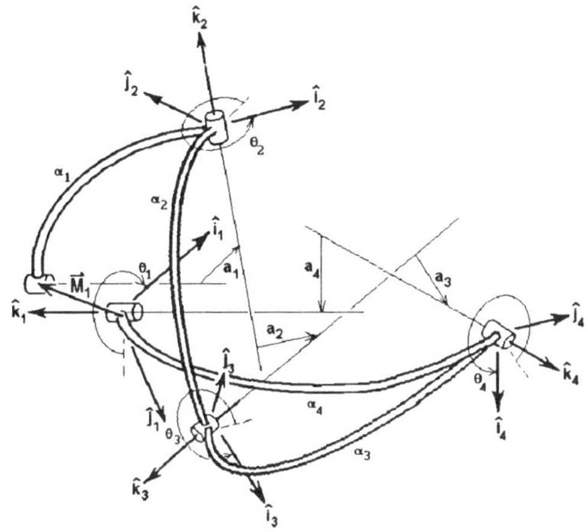

Figure 32 The properly oriented links of the mechanism are assembled so that joints 2, 3 and 4 are connected. The unknown translational displacements at these joints are guessed as zero resulting in joint 1 being unable to close. The origin-displacement equation is then used to calculate the miss in closure in terms of the coordinate frame fixed on link 4 at its joint 1.

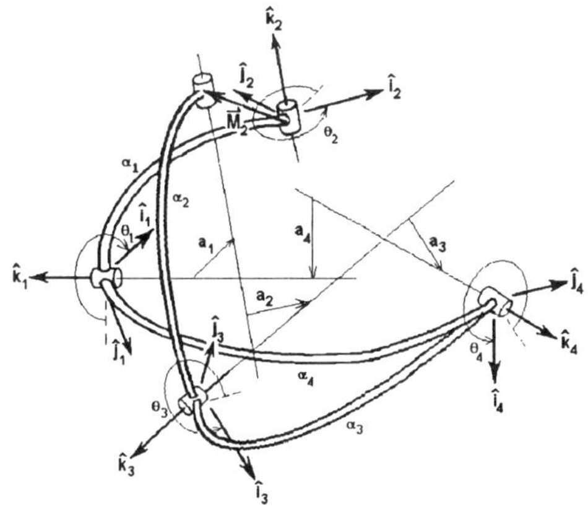

Figure 33 The properly oriented links are reassembled so that joints 1, 3 and 4 are connected. The unknown translational displacements having been guessed as zero result in a gap at joint 2. The gap, representing the miss in closure, is evaluated using the origin-displacement equation.

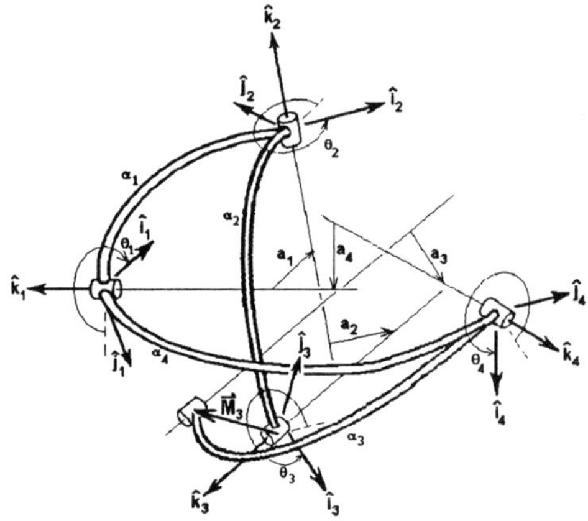

Figure 34 Now the mechanism is reassembled with joints 1, 2 and 4 connected with the links properly oriented and the translational displacements set to zero. The origin-displacement equation is used to evaluate the miss in closure at joint 3.

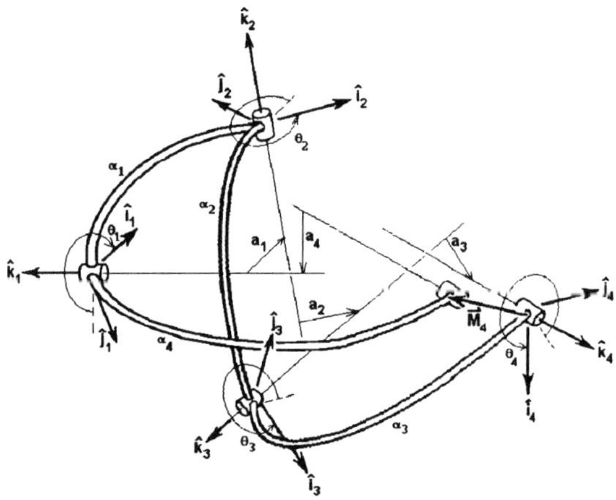

Figure 35 Finally the mechanism is assembled with joints 1, 2 and 3 connected, the links properly oriented and the translational displacements guessed as being zero. The origin-displacement equation is used to evaluate the miss at joint 4. The information about the gaps in the four assemblies can be used to estimate the corrections which must be made in the translational displacements for the mechanism to be closed.

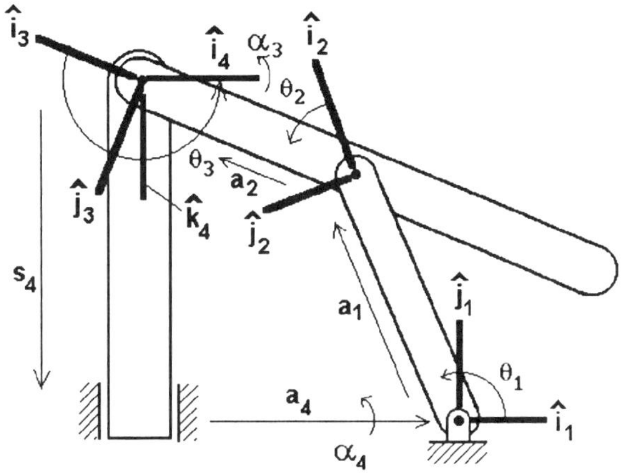

Figure 36 Hand-pump mechanism.

Exercises - Displacement Analysis

1. The mechanism shown in Figure 36 is commonly used for hand pumps. Develop an expression for displacement of the ram s_4 in terms of the displacement of the handle θ_3.

2. Develop expressions for the intermediate rotation θ_2 of a four-bar mechanism in terms of its link lengths a_1, a_2, a_3 and a_4 and the input and output rotations θ_1 and θ_4.

3. The wobble-plate mechanism is a spherical four-bar mechanism where the input crank has some twist α_1 and the other three links have a twist of $\pi/2$. Develop the input-output displacement relationship for this mechanism.

4. The Bennett mechanism is characterized by four revolute joints and dimensions such that $a_1 = a_3 = a$, $a_2 = a_4 = b$, $\alpha_1 = \alpha_3 = \alpha$, $\alpha_2 = \alpha_4 = \beta$ where $\frac{a}{\sin\alpha} = \frac{b}{\sin\beta}$. Develop the input-output displacement relationship of this mechanism for the case where $\alpha = 90°$.

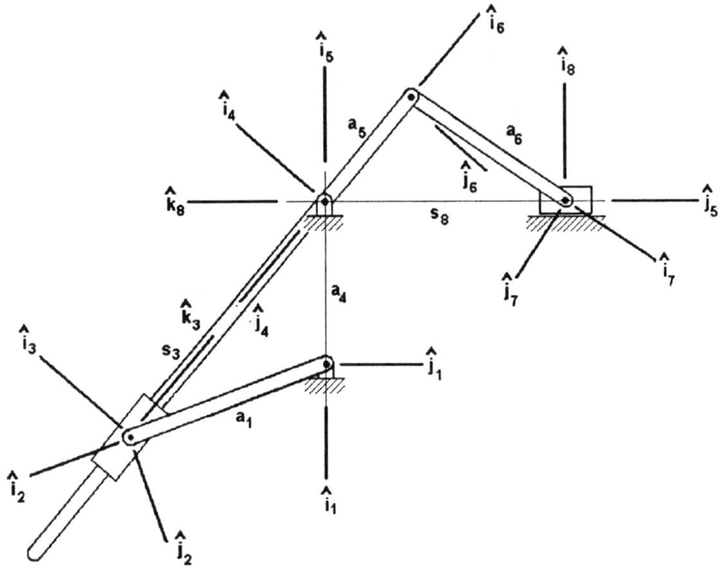

Figure 37 Whitworth quick-return

5. A spatial four-bar mechanism has links with twists $\alpha_1 = 32°$, $\alpha_2 = 54°$, $\alpha_3 = 72°$, $\alpha_4 = 135°$ and lengths $a_1 = 1$, $a_2 = 2$, $a_3 = 3$, $a_4 = 4$. The mechanism is assembled with the links properly oriented so that $\theta_1 = 42°$, $\theta_2 = 115°$, $\theta_3 = 76°$, $\theta_4 = 49°$ and the prismatic displacements s_i all set to zero so that the mechanism cannot be connected at joint 1. What is the gap at joint 1 expressed as a vector in terms of frame {1}?

6. Consider the Whitworth quick-return mechanism shown in Figure 37. It is a two-loop mechanism. Make up a table of its Denavit-Hartenberg parameters. Develop the input-output relationship by writing a condition-of-loop-closure equation for each loop and solving them simultaneously.

Chapter 5 - Velocity Analysis

Basic Concepts

Dual Velocity of Points on the Screw Axis

It has been shown that the displacement of a rigid body from one situation to another constitutes a screw motion, the simultaneous translation along and rotation about a screw axis.

The speeds at which these motions occur are

V = linear speed along the screw axis

Ω = angular speed about the screw axis

The direction and location of the screw axis can be specified by the unit screw vector

$$\hat{u} = \vec{u} + \varepsilon\left(\vec{OP} \times \vec{u}\right) \quad (355)$$

where symbol $\vec{u}$ represents a unit vector ($|\vec{u}| = 1$) and vector $\vec{OP}$ extends form the origin of the coordinate system to any point P on the screw axis.

These quantities can be combined into a "motor," the dual multiple of a unit line vector

$$\hat{V} = \left(\Omega + \varepsilon V\right)\hat{u} \quad (356)$$

It is observed that $\hat{V}$ and $\hat{u}$ are dual vectors while Ω and V are scalars.

Eq.(356) provides the dual velocity of any point P on the screw axis (all points on the screw axis have the same dual velocity).

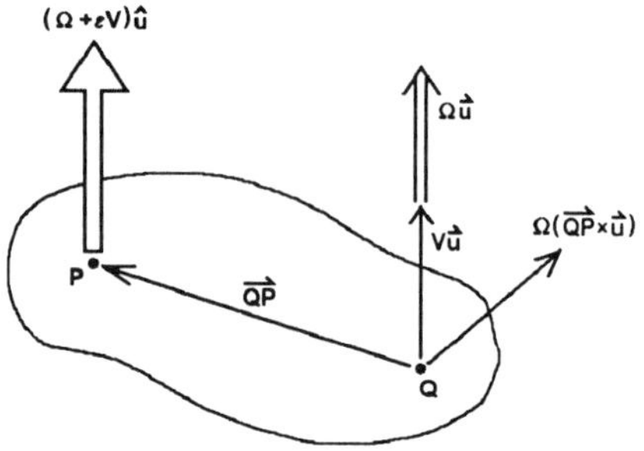

Figure 38 Dual velocities of points P and Q of a rigid body.

Dual Velocity of Points Not on the Screw Axis

Regarding the rigid body shown in Figure 38, consider point P on the screw axis as the origin of a frame and point Q as the origin of another frame with the same orientation.

The equation of transformation yields

$$^{Q}\hat{u} = \vec{u} + \varepsilon\left(\vec{QP} \times \vec{u}\right) \tag{357}$$

Substitute eq.(357) into eq.(356) and obtain the dual velocity at point Q

$$\hat{V}^{Q} = \left(\Omega + \varepsilon V\right)\left[\vec{u} + \varepsilon\left(\vec{QP} \times \vec{u}\right)\right] \tag{358}$$

$$\hat{V}^{Q} = \Omega\vec{u} + \varepsilon\left[V\vec{u} + \Omega\left(\vec{QP} \times \vec{u}\right)\right] \tag{359}$$

where the primary component $\Omega\vec{u}$ represents the angular velocity of the body and the dual component $\left[V\vec{u} + \Omega\left(\vec{QP} \times \vec{u}\right)\right]$ represents the linear velocity of point Q in the body. That linear velocity consists of a sliding velocity $V\vec{u}$ representing speed component along the screw axis and a tangential velocity $\Omega\left(\vec{QP} \times \vec{u}\right)$ directed perpendicular to the screw axis and the vector between the point Q and the screw axis.

It can be seen in eq.(359) that the angular velocity of a rigid body is the same at all points of the body.

The linear velocity of points in a rigid body consists of a sliding component in the direction of the screw axis which is constant for all points of the body and a tangential component which varies with the distance of the point from the screw axis.

On the screw axis, the tangential component of velocity vanishes. Therefore, the screw axis is unique in that the angular and linear velocities of its points are oriented along the screw axis.

An alternative form of eq.(359) is

$$\hat{\mathrm{V}}^{Q} = \Omega\vec{\mathrm{u}} + \varepsilon\left[V\vec{\mathrm{u}} + \Omega\left(\vec{\mathrm{u}} \times \vec{\mathrm{PQ}}\right)\right] \tag{360}$$

Eqs.(359) and (360) are often condensed to

$$\hat{\mathrm{V}}^{Q} = \vec{\Omega} + \varepsilon\vec{\mathrm{V}}^{Q} \tag{361}$$

From eqs.(359) and (361) we observe that

$$\Omega = \left|\vec{\Omega}\right| \tag{362}$$

The linear velocity of a point on the screw axis is

$$V = \frac{\vec{\Omega}\cdot\vec{\mathrm{V}}^{Q}}{\Omega} \tag{363}$$

This can be substantiated by substitution of the primary and dual components of eq.(361) as given in eq.(360)

$$V = \frac{\Omega\vec{\mathrm{u}}\cdot\left[V\vec{\mathrm{u}} + \Omega\left(\vec{\mathrm{QP}} \times \vec{\mathrm{u}}\right)\right]}{\Omega} \tag{364}$$

$$= \frac{\Omega\vec{\mathrm{u}}\cdot\vec{\mathrm{u}} + \Omega^{2}\vec{\mathrm{u}}\cdot\left(\vec{\mathrm{QP}} \times \vec{\mathrm{u}}\right)}{\Omega} \tag{365}$$

Now,

$$\vec{\mathrm{u}}\cdot\vec{\mathrm{u}} = 1 \tag{366}$$

and

$$\vec{\mathrm{u}}\cdot\left(\vec{\mathrm{QP}} \times \vec{\mathrm{u}}\right) = 0 \tag{367}$$

because $\left(\vec{\mathrm{QP}} \times \vec{\mathrm{u}}\right) \perp \vec{\mathrm{u}}$.

So,

$$V = \frac{\Omega V}{\Omega} = V \tag{368}$$

and eq.(363) is valid.

It is observed that the velocity along the screw axis $\vec{V}$ is independent of the point Q employed in eq.(363).

Screw Axis of a Dual Velocity

The unit screw is specified by

$$^{Q}\hat{u} = \frac{\vec{\Omega}}{\Omega} + \varepsilon\left[\frac{\vec{\Omega}\times\left(\vec{\Omega}\times\vec{V}\right)}{\Omega^{3}}\right] \tag{369}$$

which is substantiated as follows.

The primary component of unit screw $^{Q}\hat{u}$ is direction vector $\vec{u}$, which can be obtained from consideration of the primary components in eqs.(359) and (361) as

$$\vec{u} = \frac{\vec{\Omega}}{\Omega} \tag{370}$$

Now let us construct the dual component of eq.(369).

The velocity of point Q is

$$\vec{V}^{Q} = V\vec{u} + \Omega\left(\vec{QP}\times\vec{u}\right) \tag{371}$$

Substituting eqs.(370) and (371) yields

$$^{Q}\vec{u} = \frac{\Omega\vec{u}\times\left\{\Omega\vec{u}\times\left[V\vec{u}+\Omega\left(\vec{QP}\times\vec{u}\right)\right]\right\}}{\Omega^{3}} \tag{372}$$

$$= \frac{\Omega\vec{u}\times\left\{\Omega V\vec{u}\times\vec{u}+\Omega^{2}\left[\vec{u}\times\left(\vec{QP}\times\vec{u}\right)\right]\right\}}{\Omega^{3}} \tag{373}$$

Since

$$\vec{u}\times\vec{u} = 0 \tag{374}$$

and

$$\vec{u}\times\left(\vec{QP}\times\vec{u}\right) = \vec{PQ} \tag{375}$$

since $\vec{u}\perp\left(\vec{QP}\times\vec{u}\right)$.

Then

$$ {}^{Q}\vec{u} = \frac{\Omega\vec{u} \times \Omega^2 \vec{PQ}}{\Omega^3} \tag{376} $$

$$ = \frac{\Omega^3 \left(\vec{u} \times \vec{PQ}\right)}{\Omega^3} \tag{377} $$

$$ {}^{Q}\vec{u} = \vec{u} \times \vec{PQ} \tag{378} $$

or

$$ {}^{Q}\vec{u} = \vec{QP} \times \vec{u} \tag{379} $$

which is compatible with the dual component of the unit screw as specified in eq.(357).

Vector Locating the Screw Axis Given Dual Velocity at a Point

Given the dual velocity of a body at a point Q, let us locate the screw axis by determining vector $\vec{r}$ which runs from point Q to the screw axis to which it is perpendicular.

The dual velocity of the body at point Q is

$$ \hat{V}^{Q} = \vec{\Omega} + \varepsilon \vec{V}^{Q} = \Omega\vec{u} + \varepsilon\left[V\vec{u} + \Omega\left(\vec{r} \times \vec{u}\right)\right] \tag{380} $$

Now, the dual component of this dual velocity is

$$ \vec{V}^{Q} = V\vec{u} + \Omega\left(\vec{r} \times \vec{u}\right) \tag{381} $$

and so

$$ \vec{V}^{Q} \times \vec{u} = \left[V\vec{u} + \Omega\left(\vec{r} \times \vec{u}\right)\right] \times \vec{u} \tag{382} $$

$$ = \Omega\left(\vec{r} \times \vec{u}\right) \times \vec{u} \tag{383} $$

Employing a vector identity,

$$ = -\Omega\vec{r} \tag{384} $$

Substituting eq.(370) and rearranging

$$ -\vec{r} = \frac{\vec{V}^{Q} \times \vec{\Omega}}{\Omega^2} \tag{385} $$

$$ \vec{r} = \frac{\vec{\Omega} \times \vec{V}^{Q}}{\Omega^2} \tag{386} $$

Transformation-Matrix Methods

It is observed that eq.(369) is analogous to the equation of transformation which was developed to change the specification of a dual vector from one coordinate system to another with the same orientation.

Thus we realize that if we know the angular and linear velocities of a point on the screw axis of a rigid body, we can determine the angular and linear velocities of any point on the body by expressing the screw-axis dual velocity in terms of a coordinate frame with origin at the point of interest.

Just as we have used transformation matrices instead of the equation of transformation to describe dual vectors in different frames, we can use transformation matrices to determine the dual velocity of any point on a body, given the dual velocity at the screw axis. The matrix approach can take into account not only the translation between the frame with origin on the screw axis and frame with origin at the point of interest, but also any rotation between the frames.

Therefore, we adopt a notation for dual velocity such that

${}^{R}\hat{V}_{ij}^{P}$ = dual velocity at point P of body i relative to body j in terms of the unit vectors of frame {R}.

Velocities at Different Points in a Rigid Body

If we know the dual velocity ${}^{S}\hat{V}_{ij}^{S}$ of a rigid body expressed in a frame {S} with origin at a point S on the screw axis, then the dual velocity at any other point P in the body expressed in the unit vectors of a frame {P} with origin at point P and possibly rotated relative to frame {S} is

$$ {}^{P}\hat{V}_{ij}^{P} = {}^{P}_{S}\hat{T}^{S}\,\hat{V}_{ij}^{S} \tag{387} $$

Velocities Expressed in Different Frames

To express this velocity in the unit vectors of a frame {Q} different than frame {P}, it is necessary to consider only the rotation between the two frames. Therefore

$$ {}^{Q}\hat{V}_{ij}^{P} = {}^{Q}_{P}\mathrm{T}_{\mathrm{p}}\,{}^{P}\hat{V}_{ij}^{P} \tag{388} $$

where matrix ${}^{Q}_{P}\mathrm{T}_{\mathrm{p}}$ is the primary component of the dual-number coordinate-transformation matrix ${}^{Q}_{P}\hat{\mathrm{T}}$.

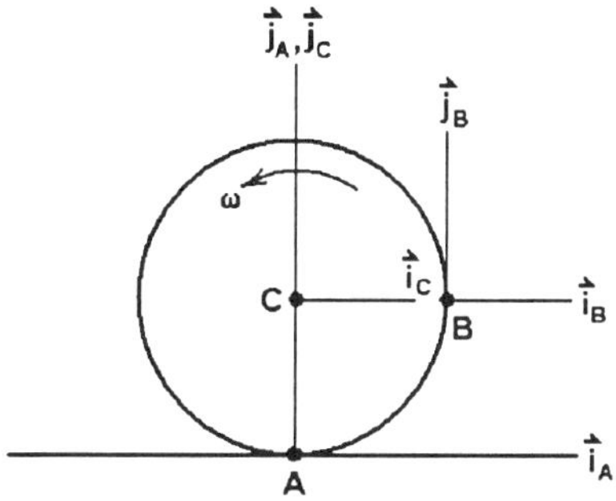

Figure 39 Velocities of points on a wheel rolling on a rail expressed in similarly oriented frames.

Because dual velocity is not a line vector, the translation between different frames in which it is expressed is not a factor.

Example

Imagine a wheel W of radius r rolling at speed ω on a rail R as shown in Figure 39. All coordinate frames have the orientations shown.

The instant center of the wheel, its screw axis, is its contact point with the rail, point A. While the wheel is in rotation there is no translation along the screw axis which is line vector $\hat{k}_A$. Therefore

$$ {}^{A}\hat{V}^{A}_{WR} = (\omega + \varepsilon 0)\hat{k}_A = \omega\hat{k}_A $$

Let us compute the dual velocity of the center of the wheel, point C.

$$ {}^{A}\hat{V}^{C}_{WR} = \Omega\vec{u} + \varepsilon\left(V\vec{u} + \Omega\vec{u} \times \vec{AC}\right) = \omega\hat{k}_A + \varepsilon\left(0\vec{k}_A + \omega\hat{k}_A \times r\vec{j}_A\right) $$

$$ {}^{A}\hat{V}^{C}_{WR} = \omega\vec{k}_A - \varepsilon\omega r\vec{i}_A $$

This result is observed to be correct.

Now, consider the use of a transformation matrix

$$ {}^{C}_{A}\hat{T} = Y(-\varepsilon r) = \begin{bmatrix} c\hat{\theta} & 0 & s\hat{\theta} \\ 0 & 1 & 0 \\ -s\hat{\theta} & 0 & c\hat{\theta} \end{bmatrix} = \begin{bmatrix} 1 & 0 & -\varepsilon r \\ 0 & 1 & 0 \\ \varepsilon r & 0 & 1 \end{bmatrix} $$

so that

$$ {}^{C}\mathrm{V}_{WR}^{C} = {}^{C}_{A}\hat{\mathrm{T}}\,{}^{C}\mathrm{V}_{WR}^{A} = \begin{bmatrix} 1 & 0 & -\varepsilon r \\ 0 & 1 & 0 \\ \varepsilon r & 0 & 1 \end{bmatrix} \begin{Bmatrix} 0 \\ 0 \\ \omega \end{Bmatrix} $$

$$ {}^{C}\mathrm{V}_{WR}^{C} = \begin{Bmatrix} -\varepsilon\omega r \\ 0 \\ \omega \end{Bmatrix} $$

In that frames {C} and {A} have the same orientation, the expressions for ${}^{C}\mathrm{V}_{WR}^{C}$ and ${}^{A}\mathrm{V}_{WR}^{C}$ are identical.

It is noted that the above expression for ${}^{C}\mathrm{V}_{WR}^{C}$ gives the magnitude and directions of the angular and linear speeds, but not the line of action (a dual velocity is not a line vector).

Now, let us determine the velocity at a point B in a frame {B} oriented the same as frame {A}.

$$ {}^{B}_{A}\hat{\mathrm{T}} = \hat{\mathrm{Y}}(-\varepsilon r)\hat{\mathrm{X}}(-\varepsilon r) = \begin{bmatrix} 1 & 0 & -\varepsilon r \\ 0 & 1 & 0 \\ \varepsilon r & 0 & 1 \end{bmatrix} \begin{bmatrix} 1 & 0 & 0 \\ 0 & 1 & \varepsilon r \\ 0 & -\varepsilon r & 1 \end{bmatrix} = \begin{bmatrix} 1 & 0 & -\varepsilon r \\ 0 & 1 & \varepsilon r \\ \varepsilon r & 0 & 1 \end{bmatrix} $$

$$ {}^{B}\hat{\mathrm{V}}_{WR}^{B} = {}^{B}_{A}\hat{\mathrm{T}}\,{}^{A}\hat{\mathrm{V}}_{WR}^{A} = \begin{bmatrix} 1 & 0 & -\varepsilon r \\ 0 & 1 & \varepsilon r \\ \varepsilon r & 0 & 1 \end{bmatrix} \begin{Bmatrix} 0 \\ 0 \\ \omega \end{Bmatrix} $$

$$ {}^{B}\hat{\mathrm{V}}_{WR}^{B} = \begin{Bmatrix} -\varepsilon\omega r \\ \varepsilon\omega r \\ \omega \end{Bmatrix} $$

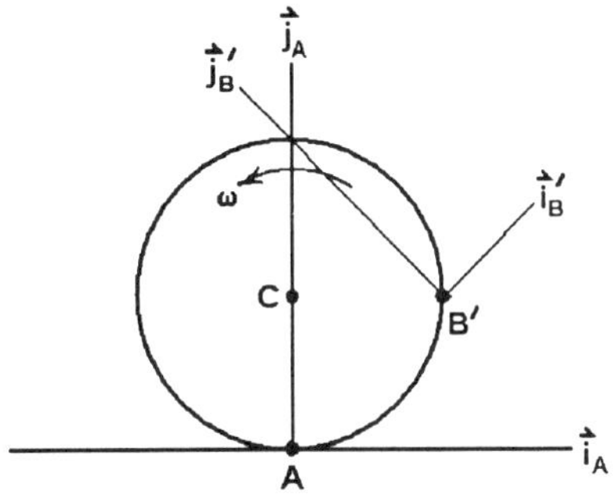

Figure 40 Velocities of points of a wheel rolling on a rail expressed in differently oriented frames.

Now, consider the case where frame {B′} is aligned as shown in Figure 40.

$$ {}^{B'}_{A}\hat{T} = \hat{Z}\left(-\frac{\pi}{4}\right)\hat{Y}(-\varepsilon r)\hat{X}(-\varepsilon r) $$

$$ = \begin{bmatrix} \sqrt{2}/2 & \sqrt{2}/2 & 0 \\ -\sqrt{2}/2 & \sqrt{2}/2 & 0 \\ 0 & 0 & 1 \end{bmatrix} \begin{bmatrix} 1 & 0 & -\varepsilon r \\ 0 & 1 & 0 \\ \varepsilon r & 0 & 1 \end{bmatrix} \begin{bmatrix} 1 & 0 & 0 \\ 0 & 1 & -\varepsilon r \\ 0 & \varepsilon r & 1 \end{bmatrix} $$

$$ {}^{B'}_{A}\hat{T} = \begin{bmatrix} \sqrt{2}/2 & \sqrt{2}/2 & 0 \\ -\sqrt{2}/2 & \sqrt{2}/2 & \varepsilon\sqrt{2}\,r \\ \varepsilon r & -\varepsilon r & 1 \end{bmatrix} $$

$$ {}^{B'}\hat{V}_{WR}^{B=B'} = {}^{B'}_{A}\hat{T}\,{}^{A}\hat{V}_{WR}^{A} = \begin{bmatrix} \sqrt{2}/2 & \sqrt{2}/2 & 0 \\ -\sqrt{2}/2 & \sqrt{2}/2 & \varepsilon\sqrt{2}\,r \\ \varepsilon r & -\varepsilon r & 1 \end{bmatrix} \begin{Bmatrix} 0 \\ 0 \\ \omega \end{Bmatrix} = \begin{Bmatrix} 0 \\ \varepsilon\sqrt{2}\,\omega r \\ \omega \end{Bmatrix} $$

The two results for the velocity of the point differ because they are expressed in different frames.

To convert the second result to the first, we apply the rotation transformation

$$ {}^{B}\hat{V}_{WR}^{B} = {}^{B}_{B'}T_{p}\,{}^{B'}\hat{V}_{WR}^{B=B'} $$

$$= \hat{Z}\left(\frac{\pi}{4}\right) {}^{B'}\hat{V}_{WR}^{B=B'} = \begin{bmatrix} 1/\sqrt{2} & -1/\sqrt{2} & 0 \\ 1/\sqrt{2} & 1/\sqrt{2} & 0 \\ 0 & 0 & 1 \end{bmatrix} \begin{Bmatrix} 0 \\ \varepsilon\sqrt{2}\omega r \\ \omega \end{Bmatrix}$$

$${}^{B}\hat{V}_{WR}^{B} = \begin{Bmatrix} -\varepsilon\omega r \\ \varepsilon\omega r \\ \omega \end{Bmatrix}$$

Relative Velocity

The relative-velocity equation of elementary kinematics can be extended to dual-number mechanics.

$$\hat{V}_{ki} = \hat{V}_{ji} + \hat{V}_{kj} \tag{389}$$

where $\hat{V}_{mn}$ denotes the dual velocity of link m relative to link n. When link n is the ground (fixed or inertial body) that subscript is often omitted in the elementary kinematics formulation.

In order to add these dual velocities, they must all represent motions at the same point and be expressed in the same frame. Let point A be the joint where links i and j are connected, and let point B be the joint where links j and k are connected as is shown in Figure 41. The relative velocities of adjacent links are usually expressed at their joint and in a frame with origin at the joint. Thus we have velocities ${}^{A}\hat{V}_{ji}^{A}$ and ${}^{B}\hat{V}_{kj}^{B}$.

To express every dual velocity at the same point and then the same unit vectors, we transform ${}^{A}\hat{V}_{ji}^{A}$ to ${}^{B}\hat{V}_{ji}^{B}$ by

$${}^{B}\hat{V}_{ji}^{B} = {}_{A}^{B}\hat{T}\, {}^{A}\hat{V}_{ji}^{A} \tag{390}$$

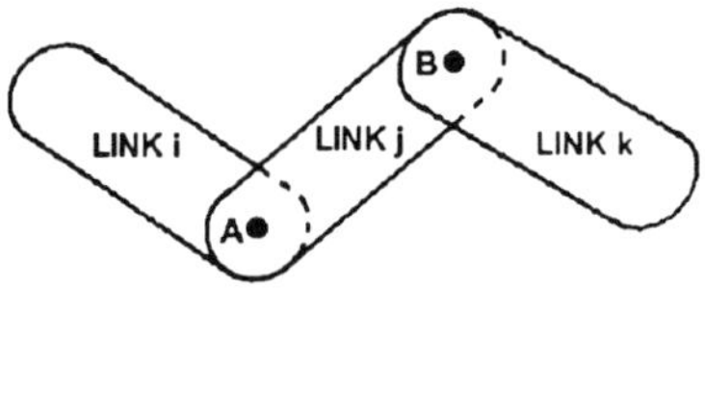

Figure 41 When several bodies move relative to one another, the dual velocities can be obtained using the dual relative-velocity equation.

Substituting eq.(390) into eq.(389) we have

$$ {}^{B}\hat{\mathrm{V}}_{ki}^{B} = {}_{A}^{B}\hat{\mathrm{T}}\;{}^{A}\hat{\mathrm{V}}_{ji}^{A} + {}^{A}\hat{\mathrm{V}}_{kj}^{b} \tag{391} $$

This is the relative dual-velocity equation.

If dual velocity ${}^{B}\hat{\mathrm{V}}_{ki}^{B}$ is required in the unit vectors of some frame {C}, we simply rotate:

$$ {}^{C}\hat{\mathrm{V}}_{ki}^{C} = {}_{B}^{C}\hat{\mathrm{T}}_{\mathrm{p}}\;{}^{B}\hat{\mathrm{V}}_{ki}^{B} \tag{392} $$

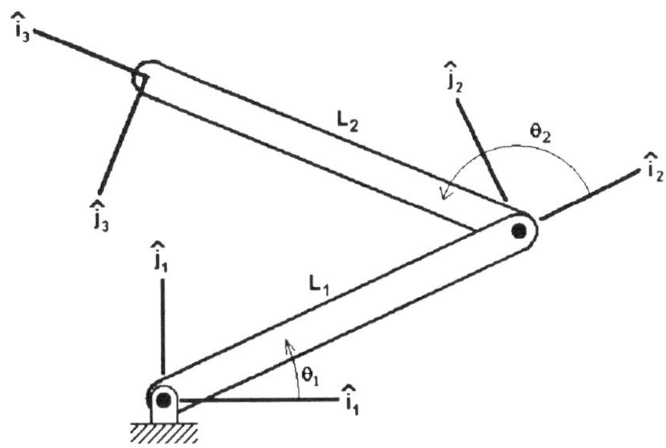

Figure 42 The two-link manipulator. Note that the ground is link 0.

Table 10 - Two-Link Manipulator				
i	θ_i	s_i	α_i	a_i
1	θ_1	0	0	L_1
2	θ_2	0	0	L_2

Example

A manipulator built out of two moving links connected by revolute joints is illustrated in Figure 42 and its Denavit-Hartenberg parameters are given in Table 10.

The joint-link transforms for this two-link manipulator are

$$ {}_2^1\hat{M} = \begin{bmatrix} c\theta_1 & -s\theta_1 & \varepsilon L_1 s\theta_1 \\ s\theta_1 & c\theta_1 & -\varepsilon L_1 c\theta_1 \\ 0 & \varepsilon L_1 & 1 \end{bmatrix} $$

$$ {}_3^2\hat{M} = \begin{bmatrix} c\theta_2 & -s\theta_2 & \varepsilon L_2 s\theta_2 \\ s\theta_2 & c\theta_2 & -\varepsilon L_2 c\theta_2 \\ 0 & \varepsilon L_2 & 1 \end{bmatrix} $$

The joints are driven by motors at velocities

$$ {}^1\hat{V}_{10}^1 = \begin{Bmatrix} 0 \\ 0 \\ \dot{\theta}_1 \end{Bmatrix} $$

$$ {}^2\hat{V}_{21}^2 = \begin{Bmatrix} 0 \\ 0 \\ \dot{\theta}_2 \end{Bmatrix} $$

These dual velocities give the speeds at which link 1 moves relative to the ground link 0 and at which link 2 moves relative to link 1.

To determine the velocity of link 2 relative to the ground link 0, the relative-velocity equation is employed:

$$ {}^2\hat{V}_{20}^2 = {}_1^2\hat{T}\,{}^1\hat{V}_{10}^1 + {}^2V_{21}^2 = \begin{bmatrix} c\theta_1 & s\theta_1 & 0 \\ -s\theta_1 & c\theta_1 & \varepsilon L_1 \\ \varepsilon L_1 s\theta_1 & -\varepsilon L_1 c\theta_1 & 1 \end{bmatrix} \begin{Bmatrix} 0 \\ 0 \\ \dot{\theta}_1 \end{Bmatrix} + \begin{Bmatrix} 0 \\ 0 \\ \dot{\theta}_2 \end{Bmatrix} $$

$$ {}^2\hat{V}_{20}^2 = \begin{Bmatrix} 0 \\ \varepsilon L_1 \dot{\theta}_1 \\ \dot{\theta}_1 + \dot{\theta}_2 \end{Bmatrix} $$

To determine the velocity of link 2 at point 3 instead of point 2,

$$ {}^3\hat{V}_{20}^3 = {}_2^3\hat{T}\,{}^2\hat{V}_{20}^2 = \begin{bmatrix} c\theta_2 & s\theta_2 & 0 \\ -s\theta_2 & c\theta_2 & \varepsilon L_2 \\ \varepsilon L_2 s\theta_2 & -\varepsilon L_2 c\theta_2 & 1 \end{bmatrix} \begin{Bmatrix} 0 \\ \varepsilon L_1 \dot{\theta}_1 \\ \dot{\theta}_1 + \dot{\theta}_2 \end{Bmatrix} $$

$$ {}^3\hat{V}_{20}^3 = \begin{Bmatrix} \varepsilon L_2 s\theta_2 \dot{\theta}_1 \\ \varepsilon\left[L_1 c\theta_2 \dot{\theta}_1 + L_2\left(\dot{\theta}_1 + \dot{\theta}_2\right)\right] \\ \dot{\theta}_1 + \dot{\theta}_2 \end{Bmatrix} $$

This dual velocity is expressed in terms of frame {3} which is fixed on the distal end of link 2 at point 3.

To express dual velocity ${}^{3}\hat{V}_{20}^{3}$ in terms of the dual vectors of frame {1} fixed on ground link 0, we apply the rotation transformation such that

$$ {}^{1}\hat{V}_{20}^{3} = {}^{1}_{3}T_{p}\,{}^{3}\hat{V}_{20}^{3} $$

where

$$ {}^{1}_{3}T_{p} = {}^{1}_{2}M_{p}\,{}^{2}_{3}M_{p} = \begin{bmatrix} c\theta_1 & -s\theta_1 & 0 \\ s\theta_1 & c\theta_1 & 0 \\ 0 & 0 & 1 \end{bmatrix} \begin{bmatrix} c\theta_2 & -s\theta_2 & 0 \\ s\theta_2 & c\theta_2 & 0 \\ 0 & 0 & 1 \end{bmatrix} = \begin{bmatrix} c\theta_{12} & -s\theta_{12} & 0 \\ s\theta_{12} & c\theta_{12} & 0 \\ 0 & 0 & 1 \end{bmatrix} $$

where $s\theta_{12} = \sin(\theta_1+\theta_2)$ and $c\theta_{12} = \cos(\theta_1+\theta_2)$.

Therefore,

$$ {}^{1}\hat{V}_{20}^{3} = \begin{bmatrix} c\theta_{12} & -s\theta_{12} & 0 \\ s\theta_{12} & c\theta_{12} & 0 \\ 0 & 0 & 1 \end{bmatrix} \begin{Bmatrix} \varepsilon L_2\, s\theta_2 \dot{\theta}_1 \\ \varepsilon\left[L_1\, c\theta_2 \dot{\theta}_1 + L_2\left(\dot{\theta}_1 + \dot{\theta}_2\right)\right] \\ \dot{\theta}_1 + \dot{\theta}_2 \end{Bmatrix} $$

$$ = \begin{Bmatrix} \varepsilon\left[L_1\, s\theta_1 \dot{\theta}_1 - L_2\, s\theta_{12}\left(\dot{\theta}_1 + \dot{\theta}_2\right)\right] \\ \varepsilon\left[L_1\, s\theta_1 \dot{\theta}_1 + L_2 c\theta_{12}\left(\dot{\theta}_1 + \dot{\theta}_2\right)\right] \\ \dot{\theta}_1 + \dot{\theta}_2 \end{Bmatrix} $$

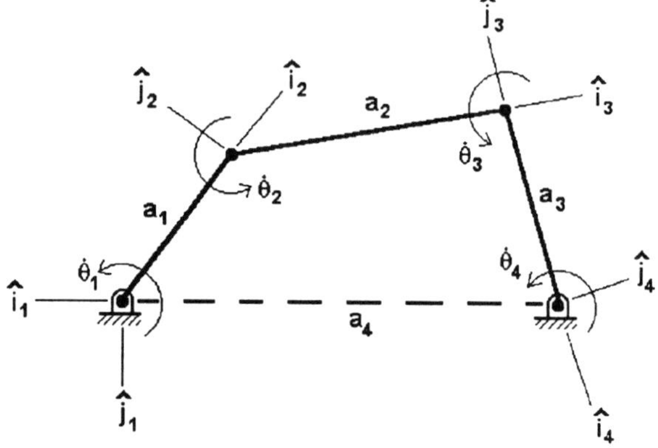

Figure 43 Velocities of the planar four-bar mechanism.

Planar Four-Bar Mechanism

The planar four-bar mechanism is considered to be operated within a single plane so that the joint offsets, s_i, are zero and the joint axes are parallel so that link twists, α_i, are zero. This linkage is illustrated in Figure 43. Based on the Denavit-Hartenberg parameters, the four joint-link transforms have the form

$$ {}_{i+1}^{\;\;i}\hat{M} = \begin{bmatrix} c\theta_i & -s\theta_i & \varepsilon a_i s\theta_i \\ s\theta_i & c\theta_i & -\varepsilon a_i c\theta_i \\ 0 & \varepsilon a_i & 1 \end{bmatrix} \tag{393} $$

Given the input-crank velocity

$$ {}^1\hat{V}_{14}^1 = \begin{Bmatrix} 0 \\ 0 \\ \dot{\theta}_1 \end{Bmatrix} \tag{394} $$

it is desired to determine the other velocities ${}^1\hat{V}_{14}^2$, ${}^1\hat{V}_{24}^2$, ${}^1\hat{V}_{24}^3$, ${}^1\hat{V}_{34}^3$ and ${}^1\hat{V}_{34}^4$

Now, the joint velocities are

$$ {}^{2}\hat{V}_{21}^{2} = \begin{Bmatrix} 0 \\ 0 \\ \dot{\theta}_{2} \end{Bmatrix} \tag{395} $$

$$ {}^{3}\hat{V}_{32}^{3} = \begin{Bmatrix} 0 \\ 0 \\ \dot{\theta}_{3} \end{Bmatrix} \tag{396} $$

$$ {}^{4}\hat{V}_{43}^{4} = \begin{Bmatrix} 0 \\ 0 \\ \dot{\theta}_{4} \end{Bmatrix} \tag{397} $$

Consider the velocity of different points on link 1,

$$ {}^{2}\hat{V}_{14}^{2} = {}_{1}^{2}\hat{M}\,{}^{1}\hat{V}_{14}^{1} = \begin{bmatrix} c\theta_{1} & s\theta_{1} & 0 \\ -s\theta_{1} & c\theta_{1} & \varepsilon a_{1} \\ \varepsilon a_{1}\, s\theta_{1} & -\varepsilon a_{1}\, c\theta_{1} & 1 \end{bmatrix} \begin{Bmatrix} 0 \\ 0 \\ \dot{\theta}_{1} \end{Bmatrix} = \begin{Bmatrix} 0 \\ \varepsilon a_{1}\dot{\theta}_{1} \\ \dot{\theta}_{1} \end{Bmatrix} \tag{398} $$

Restate this velocity in the unit vectors of the fixed frame $\{1\}$,

$$ {}^{1}\hat{V}_{14}^{2} = {}_{2}^{1}M_{p}\,{}^{2}\hat{V}_{14}^{2} = \begin{bmatrix} c\theta_{1} & -s\theta_{1} & 0 \\ s\theta_{1} & c\theta_{1} & 0 \\ 0 & 0 & 1 \end{bmatrix} \begin{Bmatrix} 0 \\ \varepsilon a_{1}\dot{\theta}_{1} \\ \dot{\theta}_{1} \end{Bmatrix} = \begin{Bmatrix} -\varepsilon a_{1}\dot{\theta}_{1}\, s\theta_{1} \\ \varepsilon a_{1}\dot{\theta}_{1}\, c\theta_{1} \\ \dot{\theta}_{1} \end{Bmatrix} \tag{399} $$

Now, let us consider the coupler. Its velocity at its connection with the input crank is

$$ {}^{2}\hat{V}_{24}^{2} = {}^{2}\hat{V}_{14}^{2} + {}^{2}\hat{V}_{21}^{2} = \begin{Bmatrix} 0 \\ \varepsilon a_{1}\dot{\theta}_{1} \\ \dot{\theta}_{1} \end{Bmatrix} + \begin{Bmatrix} 0 \\ 0 \\ \dot{\theta}_{2} \end{Bmatrix} = \begin{Bmatrix} 0 \\ \varepsilon a_{1}\dot{\theta}_{1} \\ \dot{\theta}_{1} + \dot{\theta}_{2} \end{Bmatrix} \tag{400} $$

We restate this velocity in terms of the unit vectors of fixed frame $\{1\}$,

$$ {}^{1}\hat{V}_{24}^{2} = {}_{2}^{1}M_{p}\,{}^{2}\hat{V}_{14}^{2} = \begin{bmatrix} c\theta_{1} & -s\theta_{1} & 0 \\ s\theta_{1} & c\theta_{1} & 0 \\ 0 & 0 & 1 \end{bmatrix} \begin{Bmatrix} 0 \\ \varepsilon a_{1}\dot{\theta}_{1} \\ \dot{\theta}_{1} + \dot{\theta}_{2} \end{Bmatrix} = \begin{Bmatrix} -\varepsilon a_{1}\dot{\theta}_{1}\, s\theta_{1} \\ \varepsilon a_{1}\dot{\theta}_{1}\, c\theta_{1} \\ \dot{\theta}_{1} + \dot{\theta}_{2} \end{Bmatrix} \tag{401} $$

Now, consider the velocity of point 3 on link 2, the velocity of the output crank at its connection with the output crank.

$$ {}^{3}\hat{V}_{24}^{3} = {}_{2}^{3}\hat{M}^{2}\hat{V}_{24}^{2} = \begin{bmatrix} c\theta_2 & s\theta_2 & 0 \\ -s\theta_2 & c\theta_2 & \varepsilon a_2 \\ \varepsilon a_2\, s\theta_2 & -\varepsilon a_2\, c\theta_2 & 1 \end{bmatrix} \begin{Bmatrix} 0 \\ \varepsilon a_1 \dot{\theta}_1 \\ \dot{\theta}_1 + \dot{\theta}_2 \end{Bmatrix} \tag{402} $$

$$ {}^{3}\hat{V}_{24}^{3} = \begin{Bmatrix} \varepsilon a_1 \dot{\theta}_1\, s\theta_2 \\ \varepsilon\left[a_1\dot{\theta}_1\, c\theta_2 + a_2\left(\dot{\theta}_1 + \dot{\theta}_2\right)\right] \\ \dot{\theta}_1 + \dot{\theta}_2 \end{Bmatrix} \tag{403} $$

We restate this velocity in the unit vectors of frame {1} fixed on the ground.

$$ \hat{V}_{24}^{3} = {}_{3}^{1}M_{p}\, {}^{3}\hat{V}_{24}^{3} = {}_{2}^{1}M_{p}\, {}_{3}^{2}M_{p}\, {}^{3}\hat{V}_{24}^{3} \tag{404} $$

$$ = \begin{bmatrix} c\theta_1 & -s\theta_1 & 0 \\ s\theta_1 & c\theta_1 & 0 \\ 0 & 0 & 1 \end{bmatrix} \begin{bmatrix} c\theta_2 & -s\theta_2 & 0 \\ s\theta_2 & c\theta_2 & 0 \\ 0 & 0 & 1 \end{bmatrix} \begin{Bmatrix} \varepsilon a_1 \dot{\theta}_1\, s\theta_2 \\ \varepsilon\left[a_1\dot{\theta}_1\, c\theta_2 + a_2\left(\dot{\theta}_1 + \dot{\theta}_2\right)\right] \\ \dot{\theta}_1 + \dot{\theta}_2 \end{Bmatrix} \tag{405} $$

$$ {}^{1}\hat{V}_{24}^{3} = \begin{Bmatrix} \varepsilon\left[a_1\dot{\theta}_1\, s\theta_2\, c\theta_{12} - a_1\dot{\theta}_1\, c\theta_2\, s\theta_{12} - a_2\left(\dot{\theta}_1 + \dot{\theta}_2\right) s\theta_{12}\right] \\ \varepsilon\left[a_1\dot{\theta}_1\, s\theta_2\, s\theta_{12} - a_1\dot{\theta}_1\, c\theta_2\, c\theta_{12} - a_2\left(\dot{\theta}_1 + \dot{\theta}_2\right) c\theta_{12}\right] \\ \dot{\theta}_1 + \dot{\theta}_2 \end{Bmatrix} \tag{406} $$

Now consider the velocity of the output link. The relative-velocity equation will be used to calculate the velocity of this link at point 3, its connection with the coupler.

$$ {}^{3}\hat{V}_{34}^{3} = {}^{3}\hat{V}_{24}^{3} + {}^{3}\hat{V}_{32}^{3} \tag{407} $$

$$ {}^{3}\hat{V}_{34}^{3} = \begin{Bmatrix} \varepsilon a_1 \dot{\theta}_1\, s\theta_2 \\ \varepsilon\left[a_1\dot{\theta}_1\, c\theta_2 + a_2\left(\dot{\theta}_1 + \dot{\theta}_2\right)\right] \\ \dot{\theta}_1 + \dot{\theta}_2 \end{Bmatrix} + \begin{Bmatrix} 0 \\ 0 \\ \dot{\theta}_3 \end{Bmatrix} = \begin{Bmatrix} \varepsilon a_1 \dot{\theta}_1\, s\theta_2 \\ \varepsilon\left[a_1\dot{\theta}_1\, c\theta_2 + a_2\left(\dot{\theta}_1 + \dot{\theta}_2\right)\right] \\ \dot{\theta}_1 + \dot{\theta}_2 + \dot{\theta}_3 \end{Bmatrix} \tag{408} $$

This we restate in the terms of frame {1},

$$ {}^{1}\hat{V}_{34}^{3} = {}_{3}^{1}M_{p}\, {}^{3}\hat{V}_{34}^{3} = {}_{2}^{1}M_{p}\, {}_{3}^{2}M_{p}\, {}^{3}\hat{V}_{34}^{3} \tag{409} $$

$$ {}^{1} = \begin{bmatrix} c\theta_1 & -s\theta_1 & 0 \\ s\theta_1 & c\theta_1 & 0 \\ 0 & 0 & 1 \end{bmatrix} \begin{bmatrix} c\theta_2 & -s\theta_2 & 0 \\ s\theta_2 & c\theta_2 & 0 \\ 0 & 0 & 1 \end{bmatrix} \begin{Bmatrix} \varepsilon a_1 \dot{\theta}_1\, s\theta_2 \\ \varepsilon\left[a_1\dot{\theta}_1\, c\theta_2 + a_2\left(\dot{\theta}_1 + \dot{\theta}_2\right)\right] \\ \dot{\theta}_1 + \dot{\theta}_2 + \dot{\theta}_3 \end{Bmatrix} \tag{410} $$

$$ {}^{1}\hat{V}_{34}^{3} = \begin{Bmatrix} \varepsilon\left[a_1\dot{\theta}_1\, s\theta_2\, c\theta_{12} - a_1\dot{\theta}_1\, c\theta_2\, s\theta_{12} - a_2\left(\dot{\theta}_1 + \dot{\theta}_2\right)s\theta_{12}\right] \\ \varepsilon\left[a_1\dot{\theta}_1\, s\theta_2 s\theta_{12} + a_1\dot{\theta}_1\, c\theta_2\, c\theta_{12} + a_2\left(\dot{\theta}_1 + \dot{\theta}_2\right)c\theta_{12}\right] \\ \dot{\theta}_1 + \dot{\theta}_2 + \dot{\theta}_3 \end{Bmatrix} \tag{411} $$

The velocity at point 4 on the output crank is

$$ {}^{4}\hat{V}_{34}^{4} = {}^{4}_{3}\hat{M}^{3}\hat{V}_{34}^{3} = \begin{bmatrix} c\theta_3 & s\theta_3 & 0 \\ s\theta_3 & c\theta_3 & \varepsilon a_3 \\ \varepsilon a_3 s\theta_3 & -\varepsilon a_3 c\theta_3 & 1 \end{bmatrix} \begin{Bmatrix} \varepsilon a_1\dot{\theta}_1\, s\theta_2 \\ \varepsilon\left[a_1\dot{\theta}_1\, c\theta_2 + a_2\left(\dot{\theta}_1 + \dot{\theta}_2\right)\right] \\ \dot{\theta}_1 + \dot{\theta}_2 + \dot{\theta}_3 \end{Bmatrix} \tag{412} $$

$$ {}^{4}\hat{V}_{34}^{4} = \begin{Bmatrix} \varepsilon\left[a_1\dot{\theta}_1\, s\theta_{23} + a_2\left(\dot{\theta}_1 + \dot{\theta}_2\right)s\theta_3\right] \\ \varepsilon\left[a_1\dot{\theta}_1 c\theta_{23} + a_2\left(\dot{\theta}_1 + \dot{\theta}_2\right)c\theta_3 + a_3\left(\dot{\theta}_1 + \dot{\theta}_2 + \dot{\theta}_3\right)\right] \\ \dot{\theta}_1 + \dot{\theta}_2 + \dot{\theta}_3 \end{Bmatrix} \tag{413} $$

In the unit vectors of frame {1}, this velocity is

$$ {}^{1}\hat{V}_{34}^{4} = {}^{1}_{4}M_{p}\, {}^{4}\hat{V}_{34}^{4} \tag{414} $$

$$ = \begin{bmatrix} c\theta_4 & s\theta_4 & 0 \\ -s\theta_4 & c\theta_4 & 0 \\ 0 & 0 & 1 \end{bmatrix} \begin{Bmatrix} \varepsilon\left[a_1\dot{\theta}_1\, s\theta_{23} + a_2\left(\dot{\theta}_1 + \dot{\theta}_2\right)s\theta_3\right] \\ \varepsilon\left[a_1\dot{\theta}_1\, c\theta_{23} + a_2\left(\dot{\theta}_1 + \dot{\theta}_2\right)c\theta_3 + a_3\left(\dot{\theta}_1 + \dot{\theta}_2 + \dot{\theta}_3\right)\right] \\ \dot{\theta}_1 + \dot{\theta}_2 + \dot{\theta}_3 \end{Bmatrix} \tag{415} $$

$$ {}^{1}\hat{V}_{34}^{4} = \begin{Bmatrix} \varepsilon\left[a_1\dot{\theta}_1\, s\theta_{234} + a_2\left(\dot{\theta}_1 + \dot{\theta}_2\right)s\theta_{34} + a_3\left(\dot{\theta}_1 + \dot{\theta}_2 + \dot{\theta}_3\right)s\theta_4\right] \\ \varepsilon\left[a_1\dot{\theta}_1\, c\theta_{234} + a_2\left(\dot{\theta}_1 + \dot{\theta}_2\right)c\theta_{34} + a_3\left(\dot{\theta}_1 + \dot{\theta}_2 + \dot{\theta}_3\right)c\theta_4\right] \\ \dot{\theta}_1 + \dot{\theta}_2 + \dot{\theta}_3 \end{Bmatrix} \tag{416} $$

What is the input-output velocity relationship $\dot{\theta}_4 = f\left(\dot{\theta}_1\right)$?

Since

$$ {}^{4}\hat{V}_{43}^{4} = \begin{Bmatrix} 0 \\ 0 \\ \dot{\theta}_4 \end{Bmatrix} \tag{417} $$

then

$$ {}^4\hat{V}_{34}^4 = \begin{Bmatrix} 0 \\ 0 \\ -\dot{\theta}_4 \end{Bmatrix} \tag{418} $$

this can be equated to the other expression for velocity ${}^4\hat{V}_{34}^4$ given by eq.(413).

$$ \begin{Bmatrix} 0 \\ 0 \\ -\dot{\theta}_4 \end{Bmatrix} = \begin{Bmatrix} \varepsilon\left[a_1\dot{\theta}_1 s\theta_{23} + a_2\left(\dot{\theta}_1 + \dot{\theta}_2\right)s\theta_3\right] \\ \varepsilon\left[a_1\dot{\theta}_1 c\theta_{23} + a_2\left(\dot{\theta}_1 + \dot{\theta}_2\right)c\theta_3 + a_3\left(\dot{\theta}_1 + \dot{\theta}_2 + \dot{\theta}_3\right)\right] \\ \dot{\theta}_1 + \dot{\theta}_2 + \dot{\theta}_3 \end{Bmatrix} \tag{419} $$

This equation constitutes a set of three non-trivial scalar equations which can be solved for unknown speeds $\dot{\theta}_2, \dot{\theta}_3, \dot{\theta}_4$.

$$ a_2\dot{\theta}_2\, s\theta_3 = -\left(a_1\, s\theta_{23} + a_2\, s\theta_3\right)\dot{\theta}_1 \tag{420} $$

$$ \left(a_2 c\theta_3 + a_3\right)\dot{\theta}_2 + a_3\dot{\theta}_3 = -\left(a_1\, c\theta_{23} + a_2\, c\theta_3 + a_3\right)\dot{\theta}_1 \tag{421} $$

$$ \dot{\theta}_2 + \dot{\theta}_3 + \dot{\theta}_4 = -\dot{\theta}_1 \tag{422} $$

These three equations can be combined into matrix form

$$ \begin{bmatrix} a_2\, s\theta_3 & 0 & 0 \\ a_2\, c\theta_3 + a_3 & a_3 & 0 \\ 1 & 1 & 1 \end{bmatrix} \begin{Bmatrix} \dot{\theta}_2 \\ \dot{\theta}_3 \\ \dot{\theta}_4 \end{Bmatrix} = \begin{Bmatrix} -\left(a_1\, s\theta_{23} + a_2\, s\theta_3\right)\dot{\theta}_1 \\ -\left(a_1\, c\theta_{23} + a_2\, c\theta_3 + a_3\right)\dot{\theta}_1 \\ \dot{\theta}_1 \end{Bmatrix} \tag{423} $$

which can be solved to yield

$$ \dot{\theta}_2 = -\left(\frac{a_1\, s\theta_{23}}{a_2\, s\theta_3} + 1\right)\dot{\theta}_1 \tag{424} $$

$$ \dot{\theta}_3 = \frac{a_1}{s\theta_3}\left(\frac{s\theta_{23}}{a_2} + \frac{s\theta_2}{a_3}\right)\dot{\theta}_1 \tag{425} $$

$$ \dot{\theta}_4 = -\frac{a_1\, s\theta_2}{a_3\, s\theta_3}\dot{\theta}_1 \tag{426} $$

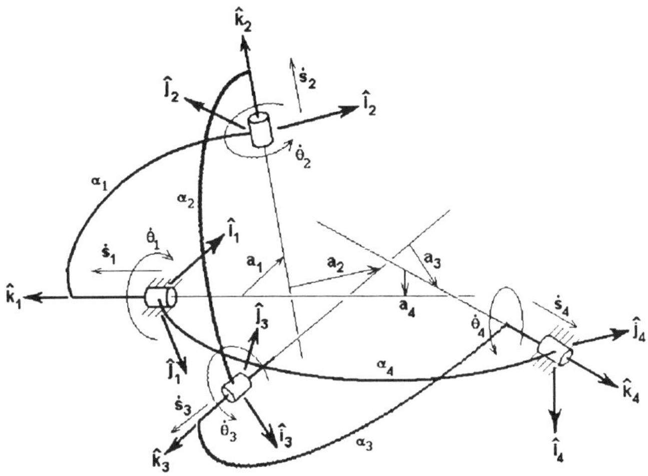

Figure 44 Velocities of the spatial four-bar mechanism.

Spatial Four-Bar Mechanism

The spatial four-bar mechanism shown in Figure 44 has an input velocity

$$^{1}V_{14}^{1} = \begin{Bmatrix} 0 \\ 0 \\ \hat{v}_1 \end{Bmatrix} \qquad (427)$$

and by definition, the other joint velocities are

$$^{2}V_{21}^{2} = \begin{Bmatrix} 0 \\ 0 \\ \hat{v}_2 \end{Bmatrix} \qquad (428)$$

$$^{3}V_{32}^{3} = \begin{Bmatrix} 0 \\ 0 \\ \hat{v}_3 \end{Bmatrix} \qquad (429)$$

$$^{4}V_{43}^{4} = \begin{Bmatrix} 0 \\ 0 \\ \hat{v}_4 \end{Bmatrix} \qquad (430)$$

where the four dual joint speeds are $\hat{v}_i = \dot{\theta}_i + \varepsilon \dot{s}_i$.

Now, the velocity of point 2 on the coupler is

$$ {}^{2}\hat{V}_{24}^{2} = {}_{1}^{2}\hat{M}^{1}\hat{V}_{14}^{1} + {}^{2}V_{21}^{2} = \begin{bmatrix} c\hat{\theta}_1 & s\hat{\theta}_1 & 0 \\ -c\hat{\alpha}_1 s\hat{\theta}_1 & c\hat{\alpha}_1 c\hat{\theta}_1 & s\hat{\alpha}_1 \\ s\hat{\alpha}_1 s\hat{\theta}_1 & -s\hat{\alpha}_1 c\hat{\theta}_1 & c\hat{\alpha}_1 \end{bmatrix} \begin{Bmatrix} 0 \\ 0 \\ \hat{v}_1 \end{Bmatrix} + \begin{Bmatrix} 0 \\ 0 \\ \hat{v}_2 \end{Bmatrix} \tag{431} $$

$$ {}^{2}\hat{V}_{24}^{2} = \begin{Bmatrix} 0 \\ s\hat{\alpha}_1 \hat{v}_1 \\ c\hat{\alpha}_1 \hat{v}_1 + \hat{v}_2 \end{Bmatrix} \tag{432} $$

The velocity of point 3 on the coupler is

$$ {}^{3}\hat{V}_{24}^{3} = {}_{2}^{3}\hat{M}^{2}\hat{V}_{24}^{2} = \begin{bmatrix} c\hat{\theta}_2 & s\hat{\theta}_2 & 0 \\ -c\hat{\alpha}_2 s\hat{\theta}_2 & c\hat{\alpha}_2 c\hat{\theta}_2 & s\hat{\alpha}_2 \\ s\hat{\alpha}_2 s\hat{\theta}_2 & -s\hat{\alpha}_2 c\hat{\theta}_2 & c\hat{\alpha}_2 \end{bmatrix} \begin{Bmatrix} 0 \\ s\hat{\alpha}_1 \hat{v}_1 \\ c\hat{\alpha}_1 \hat{v}_1 + \hat{v}_2 \end{Bmatrix} \tag{433} $$

$$ {}^{3}\hat{V}_{24}^{3} = \begin{Bmatrix} \hat{v}_1 s\hat{\alpha}_1 s\hat{\theta}_2 \\ \hat{v}_1 \left(c\hat{\alpha}_1 s\hat{\alpha}_2 + s\hat{\alpha}_1 c\hat{\alpha}_2 c\theta_2 \right) + \hat{v}_2 s\hat{\alpha}_2 \\ \hat{v}_1 \left(c\hat{\alpha}_1 c\hat{\alpha}_2 - s\hat{\alpha}_1 s\hat{\alpha}_2 c\theta_2 \right) + \hat{v}_2 c\hat{\alpha}_2 \end{Bmatrix} \tag{434} $$

Now,

$$ {}^{3}\hat{V}_{23}^{3} = -{}^{3}\hat{V}_{32}^{3} = \begin{Bmatrix} 0 \\ 0 \\ -\hat{v}_3 \end{Bmatrix} \tag{435} $$

and

$$ {}^{4}\hat{V}_{34}^{4} = -{}^{4}\hat{V}_{43}^{4} = \begin{Bmatrix} 0 \\ 0 \\ -\hat{v}_4 \end{Bmatrix} \tag{436} $$

The velocity of point 3 on the coupler can be calculated as

$$ {}^{3}\hat{V}_{24}^{3} = {}_{4}^{3}\hat{M}^{4}\hat{V}_{34}^{4} + {}^{3}\hat{V}_{23}^{3} = \begin{bmatrix} c\hat{\theta}_3 & -c\hat{\alpha}_3 s\hat{\theta}_3 & s\hat{\alpha}_3 s\hat{\theta}_3 \\ s\hat{\theta}_3 & c\hat{\alpha}_3 c\hat{\theta}_3 & -s\hat{\alpha}_3 c\hat{\theta}_3 \\ 0 & s\hat{\alpha}_3 & c\hat{\alpha}_3 \end{bmatrix} \begin{Bmatrix} 0 \\ 0 \\ -\hat{v}_4 \end{Bmatrix} + \begin{Bmatrix} 0 \\ 0 \\ -\hat{v}_3 \end{Bmatrix} \tag{437} $$

$$
{}^{3}\hat{\mathrm{V}}_{24}^{3}=\begin{Bmatrix}-\hat{v}_4\,\mathrm{s}\hat{\alpha}_3\,\mathrm{s}\hat{\theta}_3\\ \hat{v}_4\,\mathrm{s}\hat{\alpha}_3\,\mathrm{c}\hat{\theta}_3\\ -\hat{v}_4\,\mathrm{c}\hat{\alpha}_3-\hat{v}_3\end{Bmatrix} \tag{438}
$$

The two expressions for the velocity ${}^{4}\hat{\mathrm{V}}_{24}^{4}$ represented by eqs.(433) and (438) can be equated,

$$
\begin{Bmatrix}\hat{v}_1\,\mathrm{s}\hat{\alpha}_1\,\mathrm{s}\hat{\theta}_2\\ \hat{v}_1\left(\mathrm{c}\hat{\alpha}_1\,\mathrm{s}\hat{\alpha}_2+\mathrm{s}\hat{\alpha}_1\,\mathrm{c}\hat{\alpha}_2\,\mathrm{c}\theta_2\right)+\hat{v}_2\,\mathrm{s}\hat{\alpha}_2\\ \hat{v}_1\left(\mathrm{c}\hat{\alpha}_1\,\mathrm{c}\hat{\alpha}_2-\mathrm{s}\hat{\alpha}_1\,\mathrm{s}\hat{\alpha}_2\,\mathrm{c}\theta_2\right)+\hat{v}_2\,\mathrm{c}\hat{\alpha}_2\end{Bmatrix}=\begin{Bmatrix}-\hat{v}_4\,\mathrm{s}\hat{\alpha}_3\,\mathrm{s}\hat{\theta}_3\\ \hat{v}_4\,\mathrm{s}\hat{\alpha}_3\,\mathrm{c}\hat{\theta}_3\\ -\hat{v}_4\,\mathrm{c}\hat{\alpha}_3-\hat{v}_3\end{Bmatrix} \tag{439}
$$

This result constitutes three dual-number equations in the unknown dual joint velocities $\hat{v}_2$, $\hat{v}_3$ and $\hat{v}_4$ where link dimensions $\hat{\alpha}_i$, joint displacements $\hat{\theta}_i$ and input-joint velocity $\hat{v}_1$ are known. These equations can be rewritten in matrix form as

$$
\begin{bmatrix}0 & 0 & \mathrm{s}\hat{\alpha}_3\,\mathrm{s}\hat{\theta}_3\\ \mathrm{s}\hat{\alpha}_2 & 0 & -\mathrm{s}\hat{\alpha}_3\,\mathrm{c}\hat{\theta}_3\\ \mathrm{c}\hat{\alpha}_2 & 1 & \mathrm{c}\hat{\alpha}_3\end{bmatrix}\begin{Bmatrix}\hat{v}_2\\ \hat{v}_3\\ \hat{v}_4\end{Bmatrix}=\begin{Bmatrix}-\hat{v}_1\,\mathrm{s}\hat{\alpha}_1\,\mathrm{s}\hat{\theta}_2\\ -\hat{v}_1\left(\mathrm{c}\hat{\alpha}_1\,\mathrm{s}\hat{\alpha}_2+\mathrm{s}\hat{\alpha}_1\,\mathrm{c}\hat{\alpha}_2\,\mathrm{c}\theta_2\right)\\ -\hat{v}_1\left(\mathrm{c}\hat{\alpha}_1\,\mathrm{c}\hat{\alpha}_2-\mathrm{s}\hat{\alpha}_1\,\mathrm{s}\hat{\alpha}_2\,\mathrm{c}\theta_2\right)\end{Bmatrix} \tag{440}
$$

The solutions are

$$
\hat{v}_2=-\left[\frac{\left(\mathrm{c}\hat{\alpha}_1\,\mathrm{s}\hat{\alpha}_2+\mathrm{s}\hat{\alpha}_1\,\mathrm{c}\hat{\alpha}_2\,\mathrm{c}\hat{\theta}_2\right)\mathrm{s}\hat{\theta}_3+\mathrm{s}\hat{\alpha}_1\,\mathrm{s}\hat{\theta}_2\,\mathrm{c}\hat{\theta}_3}{\mathrm{s}\hat{\alpha}_2\,\mathrm{s}\hat{\theta}_3}\right]\hat{v}_1 \tag{441}
$$

$$
\hat{v}_3=\left[\frac{\left(\mathrm{s}\hat{\alpha}_2\,\mathrm{c}\hat{\alpha}_3+\mathrm{c}\hat{\alpha}_2\,\mathrm{s}\hat{\alpha}_3\,\mathrm{s}\hat{\theta}_3\right)\mathrm{s}\hat{\theta}_2+\mathrm{s}\hat{\alpha}_3\,\mathrm{c}\hat{\theta}_2\,\mathrm{s}\hat{\theta}_3}{\mathrm{s}\hat{\alpha}_2\,\mathrm{s}\hat{\alpha}_3\,\mathrm{s}\hat{\theta}_3}\right]\mathrm{s}\hat{\alpha}_1\hat{v}_1 \tag{442}
$$

$$
\hat{v}_4=-\frac{\mathrm{s}\hat{\alpha}_1\,\mathrm{s}\hat{\theta}_2}{\mathrm{s}\hat{\alpha}_3\,\mathrm{s}\hat{\theta}_3}\hat{v}_1 \tag{443}
$$

These three expressions can be expanded to yield the angular and linear speeds at the output and the two intermediate joints.

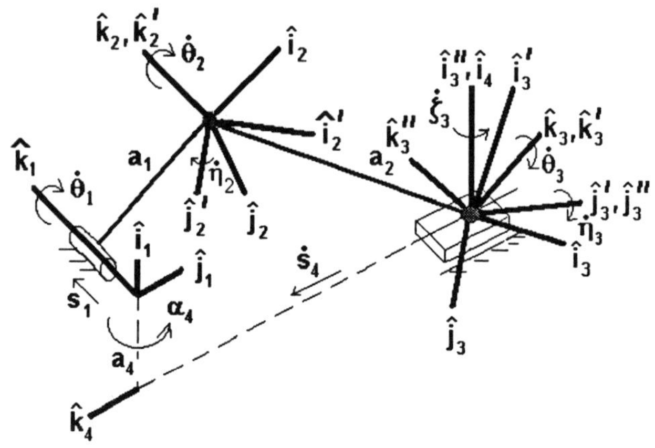

Figure 45 Velocities of the generalized slider-crank mechanism.

Generalized Slider-Crank Mechanism

The input velocity of the generalized slider-crank (CSSP) mechanism which is shown in Figure 45 is

$$ {}^{1}\hat{V}_{14}^{1} = \begin{Bmatrix} 0 \\ 0 \\ \dot{\theta}_1 + \varepsilon \dot{s}_1 \end{Bmatrix} \tag{444} $$

and by consideration of the motions allowed by the joints, the joint speeds are

$$ {}^{2}\hat{V}_{21}^{2} = \begin{Bmatrix} 0 \\ \dot{\eta}_2 \\ \dot{\theta}_2 \end{Bmatrix} \tag{445} $$

$$ {}^{3}\hat{V}_{32}^{3} = \begin{Bmatrix} \dot{\varsigma}_3 \\ \dot{\eta}_3 \\ \dot{\theta}_3 \end{Bmatrix} \tag{446} $$

$$ {}^{4}\hat{V}_{43}^{4} = \begin{Bmatrix} 0 \\ 0 \\ \varepsilon \dot{s}_4 \end{Bmatrix} \tag{447} $$

Now, the velocity of point 2 on the coupler is

$$ {}^{2}\hat{V}_{24}^{2} = {}^{2}_{1}\hat{M}^{1}\hat{V}_{14}^{1} + {}^{2}\hat{V}_{21}^{2} \tag{448} $$

$$ = \begin{bmatrix} c\theta_1 - \varepsilon s_1 s\theta_1 & s\theta_1 + \varepsilon s_1 c\theta_1 & 0 \\ -s\theta_1 - \varepsilon s_1 c\theta_1 & c\theta_1 - \varepsilon s_1 s\theta_1 & \varepsilon a_1 \\ \varepsilon a_1 s\theta_1 & -\varepsilon a_1 c\theta_1 & 1 \end{bmatrix} \begin{Bmatrix} 0 \\ 0 \\ \dot{\theta}_1 + \varepsilon \dot{s}_1 \end{Bmatrix} + \begin{Bmatrix} 0 \\ \dot{\eta}_2 \\ \dot{\theta}_2 \end{Bmatrix} \tag{449} $$

$$ {}^{2}\hat{V}_{24}^{2} = \begin{Bmatrix} 0 \\ \dot{\eta}_2 + \varepsilon \dot{\theta}_1 a_1 \\ \dot{\theta}_1 + \dot{\theta}_2 + \varepsilon \dot{s}_1 \end{Bmatrix} \tag{450} $$

The velocity of point 3 on the coupler is

$$ {}^{3}\hat{V}_{24}^{3} = {}^{3}_{2}\hat{L}^{2}\hat{V}_{24}^{2} \tag{451} $$

$$ = \begin{bmatrix} c\theta_2 c\eta_2 & s\theta_2 c\eta_2 & -s\eta_2 \\ -s\theta_2 + \varepsilon a_2 c\theta_2 s\eta_2 & c\theta_2 + \varepsilon a_2 s\theta_2 s\eta_2 & \varepsilon a_2 c\eta_2 \\ c\theta_2 s\eta_2 + \varepsilon a_2 s\theta_2 & s\theta_2 s\eta_2 - \varepsilon a_2 c\theta_2 & c\eta_2 \end{bmatrix} \begin{Bmatrix} 0 \\ \dot{\eta}_2 + \varepsilon \dot{\theta}_1 a_1 \\ \dot{\theta}_1 + \dot{\theta}_2 + \varepsilon \dot{s}_1 \end{Bmatrix} \tag{452} $$

$$ {}^{3}\hat{V}_{24}^{3} = \begin{Bmatrix} -(\dot{\theta}_1 + \dot{\theta}_2) s\eta_2 + \dot{\eta}_2 s\theta_2 c\eta_2 + \varepsilon(\dot{\theta}_1 a_1 s\theta_2 c\eta_2 - \dot{s}_1 s\eta_2) \\ \dot{\eta}_2 c\theta_2 + \varepsilon[\dot{\theta}_1(a_1 c\theta_2 + a_2 c\eta_2) + \dot{\theta}_2 a_2 c\eta_2 + \dot{\eta}_2 a_2 s\theta_2 s\eta_2] \\ (\dot{\theta}_1 + \dot{\theta}_2) c\eta_2 + \dot{\eta}_2 s\theta_2 s\eta_2 + \varepsilon(\dot{\theta}_1 a_1 s\theta_2 s\eta_2 - \dot{\eta}_2 a_2 c\theta_2 + \dot{s}_1 c\eta_2) \end{Bmatrix} \tag{453} $$

Now,

$$ {}^{3}\hat{V}_{23}^{3} = -{}^{3}\hat{V}_{32}^{3} = \begin{Bmatrix} -\dot{\varsigma}_3 \\ -\dot{\eta}_3 \\ -\dot{\theta}_3 \end{Bmatrix} \tag{454} $$

$$ {}^{4}\hat{V}_{34}^{4} = -{}^{4}\hat{V}_{43}^{4} = \begin{Bmatrix} 0 \\ 0 \\ -\varepsilon \dot{s}_4 \end{Bmatrix} \tag{455} $$

The velocity of point 1 on the coupler can also be calculated as

$$ {}^{3}\hat{V}_{24}^{3} = {}^{3}_{4}\hat{L}^{4}\hat{V}_{34}^{4} + {}^{3}\hat{V}_{23}^{3} \tag{456} $$

$$ = \begin{bmatrix} c\theta_3 c\eta_3 & -s\theta_3 c\varsigma_3 + c\theta_3 s\eta_3 s\varsigma_3 & s\theta_3 s\varsigma_3 + c\theta_3 s\eta_3 c\varsigma_3 \\ s\theta_3 c\eta_3 & c\theta_3 c\varsigma_3 + s\theta_3 s\eta_3 s\varsigma_3 & -c\theta_3 s\varsigma_3 + s\theta_3 s\eta_3 c\varsigma_3 \\ -s\eta_3 & c\eta_3 s\varsigma_3 & c\eta_3 c\varsigma_3 \end{bmatrix} \begin{Bmatrix} 0 \\ 0 \\ -\varepsilon \dot{s}_4 \end{Bmatrix} + \begin{Bmatrix} -\dot{\varsigma}_3 \\ -\dot{\eta}_3 \\ -\dot{\theta}_3 \end{Bmatrix} \tag{457} $$

$$ {}^{3}\hat{\mathrm{V}}_{24}^{3} = \left\{ \begin{matrix} -\dot{\varsigma}_3 - \varepsilon \dot{s}_4\left(\mathrm{s}\theta_3\,\mathrm{s}\varsigma_3 + \mathrm{c}\theta_3\,\mathrm{s}\eta_3\,\mathrm{c}\varsigma_3\right) \\ -\dot{\eta}_3 + \varepsilon \dot{s}_4\left(\mathrm{c}\theta_3\,\mathrm{s}\varsigma_3 - \mathrm{s}\theta_3\,\mathrm{s}\eta_3\,\mathrm{c}\varsigma_3\right) \\ -\dot{\theta}_3 - \varepsilon \dot{s}_4\,\mathrm{c}\eta_3\,\mathrm{c}\varsigma_3 \end{matrix} \right\} \tag{458} $$

The two expressions for velocity ${}^{3}\hat{\mathrm{V}}_{24}^{3}$ can be equated,

$$ \left\{ \begin{matrix} -\left(\dot{\theta}_1 + \dot{\theta}_2\right)\mathrm{s}\eta_2 + \dot{\eta}_2\,\mathrm{s}\theta_2\,\mathrm{c}\eta_2 + \varepsilon\left(\dot{\theta}_1 a_1\,\mathrm{s}\theta_2\,\mathrm{c}\eta_2 - \dot{s}_1\,\mathrm{s}\eta_2\right) \\ \dot{\eta}_2\,\mathrm{c}\theta_2 + \varepsilon\left[\dot{\theta}_1\left(a_1\,\mathrm{c}\theta_2 + a_2\,\mathrm{c}\eta_2\right) + \dot{\theta}_2 a_2\,\mathrm{c}\eta_2 + \dot{\eta}_2 a_2\,\mathrm{s}\theta_2\,\mathrm{s}\eta_2\right] \\ \left(\dot{\theta}_1 + \dot{\theta}_2\right)\mathrm{c}\eta_2 + \dot{\eta}_2\,\mathrm{s}\theta_2\,\mathrm{s}\eta_2 + \varepsilon\left(\dot{\theta}_1 a_1\,\mathrm{s}\theta_2\,\mathrm{s}\eta_2 - \dot{\eta}_2 a_2\,\mathrm{c}\theta_2 + \dot{s}_1\,\mathrm{c}\eta_2\right) \end{matrix} \right\} = \left\{ \begin{matrix} -\dot{\varsigma}_3 - \varepsilon \dot{s}_4\left(\mathrm{s}\theta_3\,\mathrm{s}\varsigma_3 + \mathrm{c}\theta_3\,\mathrm{s}\eta_3\,\mathrm{c}\varsigma_3\right) \\ -\dot{\eta}_3 + \varepsilon \dot{s}_4\left(\mathrm{c}\theta_3\,\mathrm{s}\varsigma_3 - \mathrm{s}\theta_3\,\mathrm{s}\eta_3\,\mathrm{c}\varsigma_3\right) \\ -\dot{\theta}_3 - \varepsilon \dot{s}_4\,\mathrm{c}\eta_3\,\mathrm{c}\varsigma_3 \end{matrix} \right\} \tag{459} $$

This dual-number column-matrix equation can be considered as six scalar equations

$$ -\left(\dot{\theta}_1 + \dot{\theta}_2\right)\mathrm{s}\eta_2 + \dot{\eta}_2\,\mathrm{s}\theta_2\,\mathrm{c}\eta_2 = -\dot{\varsigma}_3 \tag{460} $$

$$ \dot{\theta}_1 a_1\,\mathrm{s}\theta_2\,\mathrm{c}\eta_2 - \dot{s}_1\,\mathrm{s}\eta_2 = -\dot{s}_4\left(\mathrm{s}\theta_3\,\mathrm{s}\varsigma_3 + \mathrm{c}\theta_3\,\mathrm{s}\eta_3\,\mathrm{c}\varsigma_3\right) \tag{461} $$

$$ \dot{\eta}_2\mathrm{c}\theta_2 = -\dot{\eta}_3 \tag{462} $$

$$ \dot{\theta}_1\left(a_1\,\mathrm{c}\theta_2 + a_2\,\mathrm{c}\eta_2\right) + \dot{\theta}_2 a_2\,\mathrm{c}\eta_2 + \dot{\eta}_2 a_2\,\mathrm{s}\theta_2\,\mathrm{s}\eta_2 = \dot{s}_4\left(\mathrm{c}\theta_3\,\mathrm{s}\varsigma_3 - \mathrm{s}\theta_3\,\mathrm{s}\eta_3\,\mathrm{c}\varsigma_3\right) \tag{463} $$

$$ \left(\dot{\theta}_1 + \dot{\theta}_2\right)\mathrm{c}\eta_2 + \dot{\eta}_2\mathrm{s}\theta_2\mathrm{s}\eta_2 = -\dot{\theta}_3 \tag{464} $$

$$ \dot{\theta}_1 a_1\,\mathrm{s}\theta_2\,\mathrm{s}\eta_2 - \dot{\eta}_2 a_2\,\mathrm{c}\theta_2 + \dot{s}_1\,\mathrm{c}\eta_2 = -\dot{s}_4\,\mathrm{c}\eta_3\,\mathrm{c}\varsigma_3 \tag{465} $$

which constitute a set which can be combined in matrix form as

$$ \begin{bmatrix} -\mathrm{s}\eta_2 & 0 & \mathrm{s}\theta_2\,\mathrm{c}\eta_2 & 0 & 1 & 0 \\ 0 & 0 & 0 & 0 & 0 & \mathrm{s}\theta_3\,\mathrm{s}\varsigma_3 + \mathrm{c}\theta_3\,\mathrm{s}\eta_3\,\mathrm{c}\varsigma_3 \\ 0 & 0 & \mathrm{c}\theta_2 & 1 & 0 & 0 \\ a_2\,\mathrm{c}\eta_2 & 0 & a_2\,\mathrm{s}\theta_2\,\mathrm{s}\eta_2 & 0 & 0 & -\mathrm{c}\theta_3\,\mathrm{s}\varsigma_3 + \mathrm{s}\theta_3\,\mathrm{s}\eta_3\,\mathrm{c}\varsigma_3 \\ \mathrm{c}\eta_3 & 1 & \mathrm{s}\theta_2\,\mathrm{s}\eta_2 & 0 & 0 & 0 \\ 0 & 0 & -a_2\,\mathrm{c}\theta_2 & 0 & 0 & \mathrm{c}\eta_3\,\mathrm{c}\varsigma_3 \end{bmatrix} \left\{ \begin{matrix} \dot{\theta}_2 \\ \dot{\theta}_3 \\ \dot{\eta}_2 \\ \dot{\eta}_3 \\ \dot{\varsigma}_3 \\ \dot{s}_4 \end{matrix} \right\} = \left\{ \begin{matrix} \dot{\theta}_1\,\mathrm{s}\eta_2 \\ -\dot{\theta}_1 a_1\,\mathrm{s}\theta_2\,\mathrm{c}\eta_3 + \dot{s}_1\,\mathrm{s}\eta_2 \\ 0 \\ -\dot{\theta}_1\left(a_1\,\mathrm{c}\theta_2 + a_2\,\mathrm{c}\eta_2\right) \\ -\dot{\theta}_1\,\mathrm{c}\eta_2 \\ -\dot{\theta}_1 a_1 s\theta_2\,\mathrm{s}\eta_3 - \dot{s}_1\,\mathrm{c}\eta_2 \end{matrix} \right\} \tag{466} $$

The solutions to this equation give the joint speeds

$$\dot{\theta}_2 = \frac{\left(a_1\dot{\theta}_1\,\mathrm{s}\theta_2\,\mathrm{c}\eta_2 - \dot{s}_1\,\mathrm{s}\eta_2\right)\left(\mathrm{s}\theta_2\,\mathrm{c}\theta_3\,\mathrm{s}\eta_2\,\mathrm{c}\eta_3\,\mathrm{c}\varsigma_3 - \mathrm{c}\theta_2\,\mathrm{s}\varsigma_3\right)}{a_2\,\mathrm{c}\theta_2\,\mathrm{c}\theta_3\,\mathrm{c}\eta_2\left(\mathrm{s}\theta_3\,\mathrm{s}\varsigma_3 + \mathrm{c}\theta_3\,\mathrm{s}\eta_3\,\mathrm{c}\varsigma_3\right)} - \frac{\left(a_1\dot{\theta}_1\,\mathrm{s}\theta_2\,\mathrm{s}\eta_2 + \dot{s}_1\,\mathrm{c}\eta_2\right)\mathrm{s}\theta_2\,\mathrm{s}\eta_2}{a_2\,\mathrm{c}\theta_2\,\mathrm{c}\eta_2} + \frac{\left(a_1\dot{\theta}_1\,\mathrm{s}\theta_2\,\mathrm{c}\eta_2 - \dot{s}_1\,\mathrm{s}\eta_2\right)\mathrm{s}\theta_3}{a_2\,\mathrm{c}\theta_3\,\mathrm{c}\eta_2} - \frac{a_1\dot{\theta}_1\,\mathrm{c}\theta_2}{a_2\,\mathrm{c}\eta_2} - \dot{\theta}_1 \tag{467}$$

$$\dot{\theta}_3 = \frac{\left(a_1\dot{\theta}_1\,\mathrm{s}\theta_2\,\mathrm{c}\eta_2 - \dot{s}_1\,\mathrm{s}\eta_2\right)\left(\mathrm{c}\theta_3\,\mathrm{s}\varsigma_3 - \mathrm{s}\theta_3\,\mathrm{s}\eta_3\,\mathrm{c}\varsigma_3\right)}{a_2\left(\mathrm{s}\theta_3\,\mathrm{s}\varsigma_3 + \mathrm{c}\theta_3\,\mathrm{s}\eta_3\,\mathrm{s}\varsigma_3\right)} + \frac{a_1\dot{\theta}_1\,\mathrm{c}\theta_2}{a_2} \tag{468}$$

$$\dot{\eta}_2 = \frac{\left(\dot{s}_1\,\mathrm{s}\eta_2 - a_1\dot{\theta}_1\,\mathrm{s}\theta_2\,\mathrm{c}\eta_2\right)\mathrm{c}\eta_3\,\mathrm{c}\varsigma_3}{a_2\,\mathrm{c}\theta_2\left(\mathrm{s}\theta_3\,\mathrm{s}\varsigma_3 + \mathrm{c}\theta_3\,\mathrm{s}\eta_3\,\mathrm{c}\varsigma_3\right)} + \frac{\dot{s}_1\,\mathrm{c}\eta_2 + a_1\dot{\theta}_1\,\mathrm{s}\theta_2\,\mathrm{s}\eta_2}{a_2\,\mathrm{c}\theta_2} \tag{469}$$

$$\dot{\eta}_3 = \frac{\left(a_1\dot{\theta}_1\,\mathrm{s}\theta_2\,\mathrm{c}\eta_2 - \dot{s}_1\,\mathrm{s}\eta_2\right)\mathrm{c}\eta_3\,\mathrm{c}\varsigma_3}{a_2\left(\mathrm{s}\theta_3\,\mathrm{s}\varsigma_3 + \mathrm{c}\theta_3\,\mathrm{s}\eta_3\,\mathrm{c}\varsigma_3\right)} - \frac{\dot{s}_1\,\mathrm{c}\eta_2 + a_1\dot{\theta}_1\,\mathrm{s}\theta_2\,\mathrm{s}\eta_2}{a_2} \tag{470}$$

$$\dot{\varsigma}_3 = \frac{\left(a_1\dot{\theta}_1\,\mathrm{c}\eta_2 - \dot{s}_1\,\mathrm{s}\eta_2\right)\left(\mathrm{c}\eta_3\,\mathrm{s}\theta_2\,\mathrm{c}\theta_3\,\mathrm{c}\varsigma_3 - \mathrm{s}\eta_2\,\mathrm{c}\theta_2\,\mathrm{s}\varsigma_3\right)}{a_2\,\mathrm{c}\theta_2\,\mathrm{c}\theta_3\,\mathrm{c}\eta_2\left(\mathrm{s}\theta_3\,\mathrm{s}\varsigma_3 + \mathrm{c}\theta_3\,\mathrm{s}\eta_3\,\mathrm{c}\varsigma_3\right)} + \frac{\left(a_1\dot{\theta}_1\,\mathrm{s}\theta_2\,\mathrm{c}\eta_2 - \dot{s}_1\,\mathrm{s}\eta_2\right)\mathrm{c}\theta_2\,\mathrm{s}\theta_3\,\mathrm{s}\eta_2 + \left(a_1\dot{\theta}_1\,\mathrm{s}\eta_2 - \dot{s}_1\,\mathrm{s}\theta_2\,\mathrm{c}\eta_2\right)\mathrm{c}\theta_3}{a_2\,\mathrm{c}\theta_2\,\mathrm{c}\theta_3\,\mathrm{c}\eta_2} \tag{471}$$

$$\dot{s}_4 = \frac{\dot{s}_1\,\mathrm{s}\eta_2 - \dot{\theta}_1 a_1\,\mathrm{s}\theta_2\,\mathrm{c}\eta_2}{\mathrm{s}\theta_3\,\mathrm{s}\varsigma_3 + \mathrm{c}\theta_3\,\mathrm{s}\eta_3\,\mathrm{c}\varsigma_3} \tag{472}$$

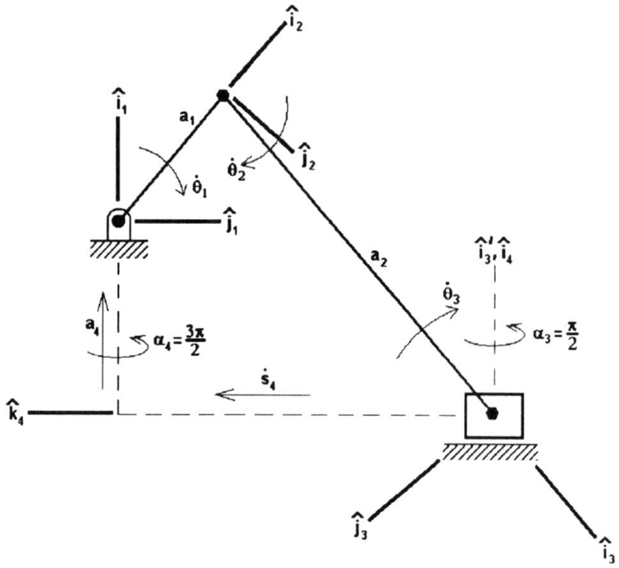

Figure 46 Velocities of the planar slider-crank mechanism.

Planar Slider-Crank Mechanism

A planar slider-crank mechanism is shown in Figure 46. The input velocity is

$$ {}^{1}\hat{V}_{14}^{1} = \begin{Bmatrix} 0 \\ 0 \\ \dot{\theta}_1 \end{Bmatrix} \tag{473}$$

and by definition the joint velocities are

$$ {}^{2}\hat{V}_{21}^{2} = \begin{Bmatrix} 0 \\ 0 \\ \dot{\theta}_2 \end{Bmatrix} \tag{474}$$

$$ {}^{3}\hat{V}_{32}^{3} = \begin{Bmatrix} 0 \\ 0 \\ \dot{\theta}_3 \end{Bmatrix} \tag{475}$$

$$ {}^{4}\hat{V}_{43}^{4} = \begin{Bmatrix} 0 \\ 0 \\ \varepsilon \dot{s}_4 \end{Bmatrix} \tag{476} $$

Now, the velocity of point 2 on the coupler is

$$ {}^{2}\hat{V}_{24}^{2} = {}_{1}^{2}\hat{M}^{1}\hat{V}_{14}^{1} + {}^{2}\hat{V}_{21}^{2} \tag{477} $$

$$ = \begin{bmatrix} c\theta_1 & s\theta_1 & 0 \\ -s\theta_1 & c\theta_1 & \varepsilon a_1 \\ \varepsilon a_1 s\theta_1 & -\varepsilon a_1 c\theta_1 & 1 \end{bmatrix} \begin{Bmatrix} 0 \\ 0 \\ \dot{\theta}_1 \end{Bmatrix} + \begin{Bmatrix} 0 \\ 0 \\ \dot{\theta}_2 \end{Bmatrix} = \begin{Bmatrix} 0 \\ \varepsilon \dot{\theta}_1 a_1 \\ \dot{\theta}_1 + \dot{\theta}_2 \end{Bmatrix} \tag{478} $$

The velocity of point 3 on the coupler is

$$ {}^{3}\hat{V}_{24}^{3} = {}_{2}^{3}\hat{M}^{2}\hat{V}_{24}^{2} = \begin{bmatrix} c\theta_2 & s\theta_2 & 0 \\ -s\theta_2 & c\theta_2 & \varepsilon a_2 \\ \varepsilon a_2 s\theta_2 & -\varepsilon a_2 c\theta_2 & 1 \end{bmatrix} \begin{Bmatrix} 0 \\ \varepsilon \dot{\theta}_1 a_1 \\ \dot{\theta}_1 + \dot{\theta}_2 \end{Bmatrix} \tag{479} $$

$$ {}^{3}\hat{V}_{24}^{3} = \begin{Bmatrix} \varepsilon \dot{\theta}_1 a_1 s\theta_2 \\ \varepsilon\left[\dot{\theta}_1\left(a_1 c\theta_2 + a_2\right) + \dot{\theta}_2 a_2\right] \\ \dot{\theta}_1 + \dot{\theta}_2 \end{Bmatrix} \tag{480} $$

Now,

$$ {}^{3}\hat{V}_{23}^{3} = -{}^{3}\hat{V}_{32}^{3} = \begin{Bmatrix} 0 \\ 0 \\ -\dot{\theta}_3 \end{Bmatrix} \tag{481} $$

$$ {}^{4}\hat{V}_{34}^{4} = -{}^{4}\hat{V}_{34}^{4} = \begin{Bmatrix} 0 \\ 0 \\ -\varepsilon \dot{s}_4 \end{Bmatrix} \tag{482} $$

The velocity of point 3 on the coupler can also be calculated as

$$ {}^{3}\hat{V}_{24}^{3} = {}_{4}^{3}\hat{M}^{4}\hat{V}_{34}^{4} + {}^{3}V_{23}^{3} = \begin{bmatrix} c\theta_3 & 0 & s\theta_3 \\ s\theta_3 & 0 & -c\theta_3 \\ 0 & 1 & 0 \end{bmatrix} \begin{Bmatrix} 0 \\ 0 \\ -\varepsilon \dot{s}_4 \end{Bmatrix} + \begin{Bmatrix} 0 \\ 0 \\ -\dot{\theta}_3 \end{Bmatrix} \tag{483} $$

$$ {}^{3}\hat{V}_{24}^{3} = \begin{Bmatrix} -\varepsilon \dot{s}_4 s\theta_3 \\ \varepsilon \dot{s}_4 c\theta_3 \\ -\dot{\theta}_3 \end{Bmatrix} \tag{484} $$

The two expressions for velocity ${}^{3}\hat{V}_{24}^{3}$ can be equated,

$$\begin{Bmatrix} \varepsilon\dot{\theta}_1 a_1 \, s\theta_2 \\ \varepsilon\left[\dot{\theta}_1\left(a_1 \, c\theta_2 + a_2\right) + \dot{\theta}_2 a_2\right] \\ \dot{\theta}_1 + \dot{\theta}_2 \end{Bmatrix} = \begin{Bmatrix} -\varepsilon\dot{s}_4 \, s\theta_3 \\ \varepsilon\dot{s}_4 \, c\theta_3 \\ -\dot{\theta}_3 \end{Bmatrix} \tag{485}$$

This dual column-matrix equation can be rewritten as a set of three non-trivial scalar equations

$$\dot{\theta}_1 a_1 \, s\theta_2 = -\dot{s}_4 \, s\theta_3 \tag{486}$$

$$\dot{\theta}_1\left(a_1 \, c\theta_2 + a_2\right) + \dot{\theta}_2 a_2 = \dot{s}_4 \, c\theta_3 \tag{487}$$

$$\dot{\theta}_1 + \dot{\theta}_2 = -\dot{\theta}_3 \tag{488}$$

which can be combined in matrix form as

$$\begin{bmatrix} 0 & 0 & -s\theta_3 \\ a_2 & 0 & c\theta_3 \\ 1 & 1 & 0 \end{bmatrix} \begin{Bmatrix} \dot{\theta}_2 \\ \dot{\theta}_3 \\ \dot{s}_4 \end{Bmatrix} = \begin{Bmatrix} \dot{\theta}_1 a_1 \, s\theta_2 \\ -\dot{\theta}_1\left(a_1 \, c\theta_2 + a_2\right) \\ -\dot{\theta}_1 \end{Bmatrix} \tag{489}$$

The solutions give the joint speeds

$$\dot{\theta}_2 = -\left(\frac{a_1 \, s\theta_{23}}{a_2 \, s\theta_3} + 1\right)\dot{\theta}_1 \tag{490}$$

$$\dot{\theta}_3 = \frac{a_1 \, s\theta_{23}}{a_2 \, s\theta_3}\dot{\theta}_1 \tag{491}$$

$$\dot{s}_4 = -\frac{a_1 \, s\theta_2}{s\theta_3}\dot{\theta}_1 \tag{492}$$

Exercises - Velocity Analysis

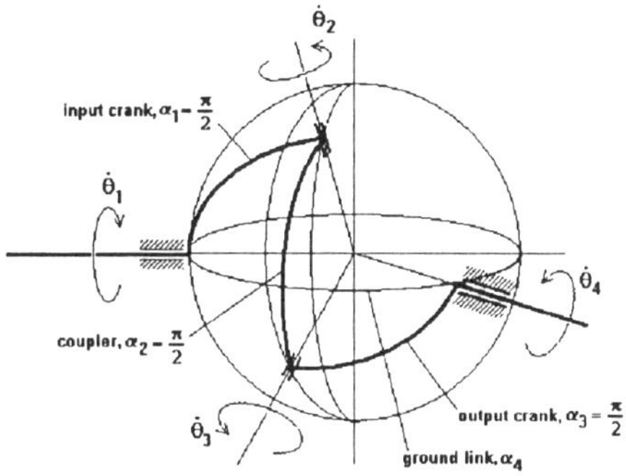

Figure 47 Velocities of the Cardan joint.

1. Develop expressions for the joint speeds of the Cardan joint, a spherical four-bar mechanism with link twists $\alpha_1=\alpha_2=\alpha_3={}^{\pi}/_2$ as shown in Figure 47.

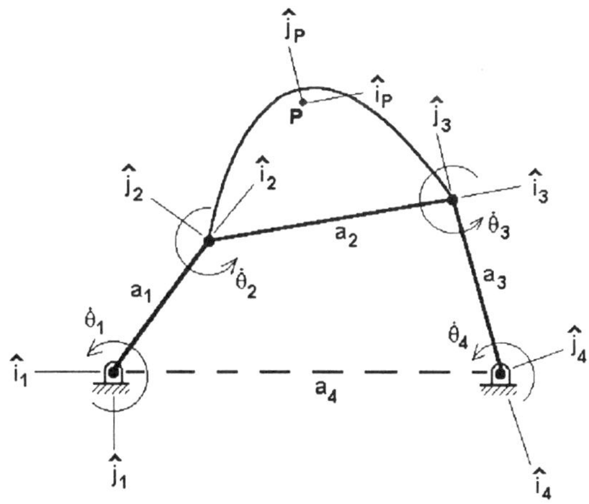

Figure 48 Four-bar mechanism with tracer point *P* on coupler link.

2. Consider a tracer point *P* fixed on the coupler of a four-bar mechanism as shown in Figure 48. This tracer point *P* is located at coordinates $\{x,y,0\}$ in the frame fixed on the distal end of the coupler. What is the absolute dual velocity of the coupler at the tracer point?

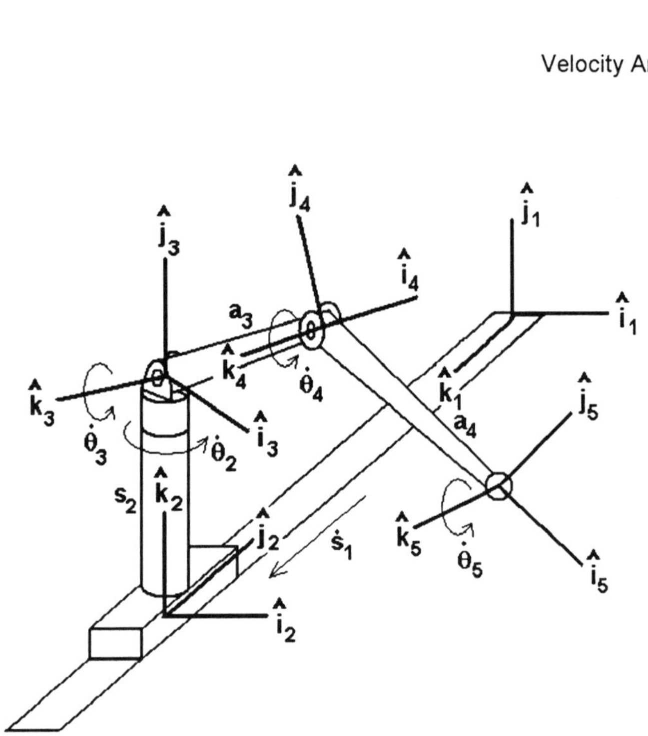

Figure 49 Robot system.

3. Consider the robot system shown in Figure 49. Given that the joints are being driven at the speeds shown, what is the absolute dual velocity of hand frame {5}?

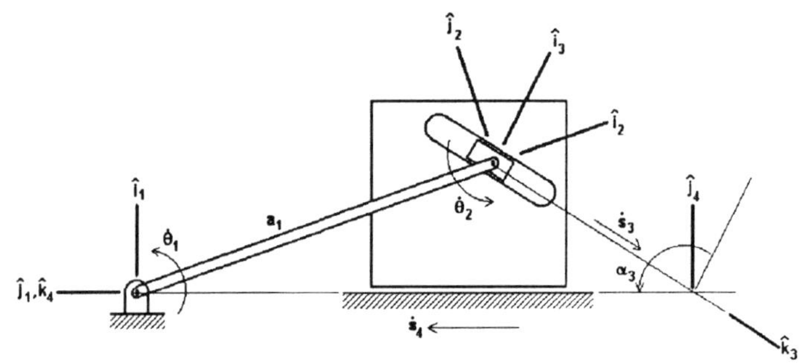

Figure 50 RRPP mechanism.

4. Consider the RRPP mechanism shown in Figure 50. Considering that the displacements are known, develop expressions for the intermediate and output joint speeds in terms of the input speed $\dot{\theta}_1$

Dual Force

If a force ${}^{A}\vec{\mathrm{F}}_{A}$ and a torque ${}^{A}\vec{\mathrm{T}}_{A}$, expressed in a coordinate frame {A} with origin at point A, act on a body at that point A, then the dual force ${}^{A}\hat{\mathrm{F}}_{A}$ at that point is defined as

$$ {}^{A}\hat{\mathrm{F}}_{A} = {}^{A}\vec{\mathrm{F}}_{A} + \varepsilon\, {}^{A}\vec{\mathrm{T}}_{A} \tag{493} $$

or

$$ {}^{A}\hat{\mathrm{F}}_{A} = \begin{Bmatrix} {}^{A}\hat{F}_{A,i} \\ {}^{A}\hat{F}_{A,j} \\ {}^{A}\hat{F}_{A,k} \end{Bmatrix} = \begin{Bmatrix} {}^{A}F_{A,i} \\ {}^{A}F_{A,j} \\ {}^{A}F_{A,k} \end{Bmatrix} + \varepsilon \begin{Bmatrix} {}^{A}T_{A,i} \\ {}^{A}T_{A,j} \\ {}^{A}T_{A,k} \end{Bmatrix} \tag{494} $$

To express the effect of this dual force at any other point in the body, we can apply a coordinate transformation just as we did to consider the velocities of different points in a body.

$$ {}^{B}\hat{\mathrm{F}}_{A} = {}^{B}_{A}\hat{\mathrm{M}}\, {}^{A}\hat{\mathrm{F}}_{A} \tag{495} $$

Example

Dual force ${}^{A}\hat{\mathrm{F}}_{A} = F_{A}\hat{\mathrm{k}} + \varepsilon T_{A}\hat{\mathrm{k}}$ acts at point A. What is a force of equivalent effect acting at point B a distance r units away in the $\hat{\mathrm{j}}$ direction?

$$ {}^{B}\hat{\mathrm{F}}_{A} = {}^{B}_{A}\hat{\mathrm{M}}\, {}^{A}\hat{\mathrm{F}}_{A} = \begin{bmatrix} 1 & 0 & \varepsilon r \\ 0 & 1 & 0 \\ \varepsilon r & 0 & 1 \end{bmatrix} \begin{Bmatrix} 0 \\ 0 \\ F_{A} + \varepsilon T_{A} \end{Bmatrix} = \begin{Bmatrix} -\varepsilon r F_{A} \\ 0 \\ F_{A} + \varepsilon T_{A} \end{Bmatrix} $$

Example

A force $\hat{\mathrm{F}} = 5\vec{\mathrm{i}} + 2\vec{\mathrm{j}} + 7\vec{\mathrm{k}}$ applied at a point on a rigid body. Replace this force with an equivalent force system at a point displaced $\vec{\mathrm{r}} = 3\vec{\mathrm{i}} + 8\vec{\mathrm{j}} + 4\vec{\mathrm{k}}$ from the original point.

$$
{}^2\hat{\mathrm{F}}_1 = {}^2_1\hat{\mathrm{M}}^1\hat{\mathrm{F}}_1 = \begin{bmatrix} 1 & \varepsilon 4 & -\varepsilon 8 \\ -\varepsilon 4 & 1 & \varepsilon 3 \\ \varepsilon 8 & -\varepsilon 3 & 1 \end{bmatrix} \begin{Bmatrix} 5 \\ 2 \\ 7 \end{Bmatrix} = \begin{Bmatrix} 5-\varepsilon 48 \\ 2+\varepsilon 1 \\ 7+\varepsilon 36 \end{Bmatrix}
$$

Equilibrium

"Equilibrium" is defined as a state where the resultant force on a particle vanishes. A rigid body is in equilibrium if each of its constituent particles is in equilibrium, and a system is in equilibrium if each of its constituent rigid bodies is in equilibrium.

A rigid body is in equilibrium if the summation of all the external dual forces applied on the rigid body, considered at the same point and in the same frame, vanish. This can be stated as

$$
\sum_{A=1}^{n} {}^B\hat{\mathrm{F}}_A = 0 \tag{496}
$$

where each of the dual forces applied at the n points A on a rigid body is considered in terms of a frame {B} with origin at a point B.

This can also be written

$$
\sum_{A=1}^{n} {}^B_A\hat{\mathrm{M}}^A\hat{\mathrm{F}}_A = 0 \tag{497}
$$

A mechanism is considered to be a system in static equilibrium if each of its links is held stationary by a set of applied dual forces.

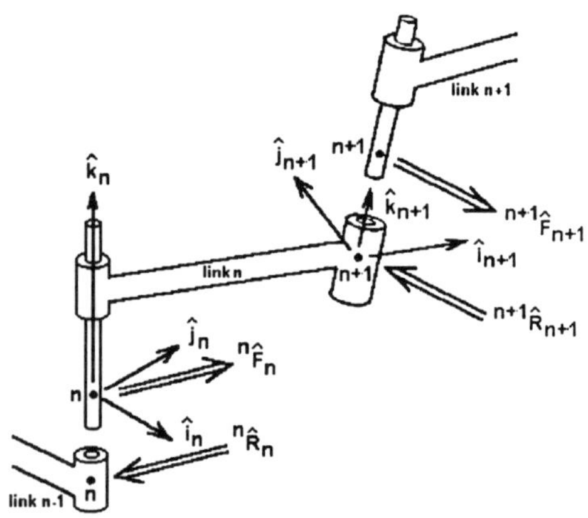

Figure 51 Static equilibrium of a general link.

Static Equilibrium of a General Link

Consider a link n connected by joints to links n-1 and n+1 and subjected to dual forces as shown in Figure 51. Link n-1 exerts dual force ${}^{n}\hat{\mathrm{F}}_{n}$ on link n at joint n. Link n exerts dual force ${}^{n+1}\hat{\mathrm{F}}_{n+1}$ on link n+1 at joint n+1.

Newton's third law indicates that link n+1 will exert a reaction dual force ${}^{n+1}\hat{\mathrm{R}}_{n+1}$ on link n such that

$$ {}^{n+1}\hat{\mathrm{R}}_{n+1} = -{}^{n+1}\hat{\mathrm{F}}_{n+1} \tag{498} $$

Consequently, neglecting gravity, there are two forces acting on link n, and if the link is in static equilibrium,

$$ 0 = {}^{n+1}_{n}\hat{\mathrm{M}}\,{}^{n}\hat{\mathrm{F}}_{n} + {}^{n+1}\hat{\mathrm{R}}_{n+1} \tag{499} $$

or

$$ 0 = {}^{n+1}_{n}\hat{\mathrm{M}}\,{}^{n}\hat{\mathrm{F}}_{n} - {}^{n+1}\hat{\mathrm{F}}_{n+1} \tag{500} $$

$$ \begin{Bmatrix} 0 \\ 0 \\ 0 \end{Bmatrix} = \begin{bmatrix} \mathrm{c}\hat{\theta}_n & \mathrm{s}\hat{\theta}_n & 0 \\ -\mathrm{c}\hat{\alpha}_n\,\mathrm{s}\hat{\theta}_n & \mathrm{c}\hat{\alpha}_n\,\mathrm{c}\hat{\theta}_n & \mathrm{s}\hat{\alpha}_n \\ \mathrm{s}\hat{\alpha}_n\,\mathrm{s}\hat{\theta}_n & -\mathrm{s}\hat{\alpha}_n\,\mathrm{c}\hat{\theta}_n & \mathrm{c}\hat{\alpha}_n \end{bmatrix} \begin{Bmatrix} {}^{n}\hat{F}_{n,i} \\ {}^{n}\hat{F}_{n,j} \\ {}^{n}\hat{F}_{n,k} \end{Bmatrix} - \begin{Bmatrix} {}^{n+1}\hat{F}_{n+1,i} \\ {}^{n+1}\hat{F}_{n+1,j} \\ {}^{n+1}\hat{F}_{n+1,k} \end{Bmatrix} \tag{501} $$

Upon expansion this becomes

$$ \begin{Bmatrix} 0 \\ 0 \\ 0 \end{Bmatrix} = \left\{ \begin{array}{c} {}^{n}F_{n,i}\,\mathrm{c}\theta_n + {}^{n}F_{n,j}\,\mathrm{s}\theta_n - {}^{n+1}F_{n+1,i} \\ +\varepsilon\left[{}^{n}T_{n,i}\,\mathrm{c}\theta_n + {}^{n}T_{n,j}\,\mathrm{s}\theta_n + {}^{n+1}T_{n+1,i} - s_n\left({}^{n}F_{n,i}\,\mathrm{s}\theta_n - {}^{n}F_{n}\,\mathrm{c}\theta_n \right) \right] \\ \\ -{}^{n}F_{n,i}\,\mathrm{c}\alpha_n\,\mathrm{s}\theta_n + {}^{n}F_{n,j}\,\mathrm{c}\alpha_n\,\mathrm{c}\theta_n + {}^{n}F_{n,k}\,\mathrm{s}\alpha_n - {}^{n+1}F_{n+1,j} \\ +\varepsilon\left[\begin{array}{c} -{}^{n}T_{n,i}\,\mathrm{c}\alpha_n\,\mathrm{s}\theta_n + {}^{n}T_{n,j}\,\mathrm{c}\alpha_n\,\mathrm{c}\theta_n + {}^{n}T_{n,k}\,\mathrm{s}\alpha_n - {}^{n+1}T_{n+1,j} \\ +a_n\left({}^{n}F_{n,i}\,\mathrm{s}\alpha_n\,\mathrm{s}\theta_n - {}^{n}F_{n,j}\,\mathrm{s}\alpha_n\,\mathrm{c}\theta_n + {}^{n}F_{n,k}\,\mathrm{c}\alpha_n \right) \\ -s_n\left({}^{n}F_{n,i}\,\mathrm{c}\alpha_n\,\mathrm{c}\theta_n + {}^{n}F_{n,j}\,\mathrm{c}\alpha_n\,\mathrm{s}\theta_n \right) \end{array} \right] \\ \\ +{}^{n}F_{n,i}\,\mathrm{s}\alpha_n\,\mathrm{s}\theta_n - {}^{n}F_{n,j}\,\mathrm{s}\alpha_n\,\mathrm{c}\theta_n + {}^{n}F_{n,k}\,\mathrm{c}\alpha_n - {}^{n+1}F_{n+1,k} \\ +\varepsilon\left[\begin{array}{c} {}^{n}T_{n,i}\,\mathrm{s}\alpha_n\,\mathrm{s}\theta_n - {}^{n}T_{n,j}\,\mathrm{s}\alpha_n\,\mathrm{c}\theta_n + {}^{n}T_{n,k}\,\mathrm{c}\alpha_n - {}^{n+1}T_{n+1,k} \\ +a_n\left({}^{n}F_{n,i}\,\mathrm{c}\alpha_n\,\mathrm{s}\theta_n - {}^{n}F_{n,j}\,\mathrm{c}\alpha_n\,\mathrm{c}\theta_n - {}^{n}F_{n,k}\,\mathrm{s}\alpha_n \right) \\ +s_n\left({}^{n}F_{n,i}\,\mathrm{s}\alpha_n\,\mathrm{c}\theta_n + {}^{n}F_{n,j}\,\mathrm{s}\alpha_n\,\mathrm{s}\theta_n \right) \end{array} \right] \end{array} \right\} \tag{502} $$

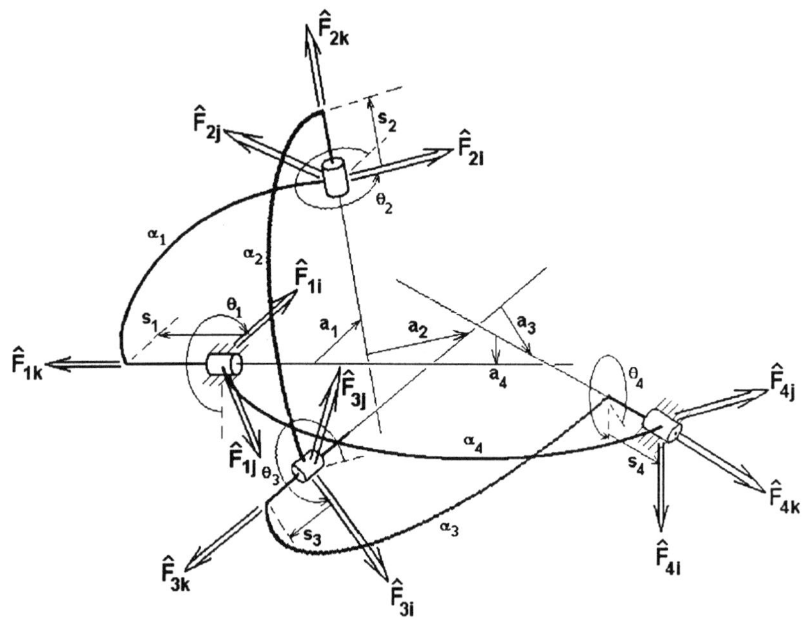

Figure 52 Dual forces acting on a spatial four-bar mechanism in static equilibrium.

Spatial Four-Bar Mechanism

The equations of equilibrium of the frictionless spatial four-bar linkage illustrated in Figure 52 are developed from eq.(502).

Under static conditions and in the absence of friction, an intermediate cylindrical joint cannot support a torque about its axis, therefore,

$$T_{2k} = 0 \tag{503}$$

and

$$T_{3k} = 0 \tag{504}$$

Substituting eqs.(503) and (504) into eq.(502) and considering each of the three moving links, we have 18 simultaneous scalar equations as follows:

$$0 = F_{1i}\,c\theta_1 + F_{1j}\,s\theta_1 - F_{2i} \tag{505}$$

$$0 = T_{1i}\,c\theta_i + T_{1j}\,s\theta_1 - T_{2i} - s_1\left(F_{1i}\,s\theta_1 - F_{1j}\,c\theta_1\right) \tag{506}$$

$$0 = -\,F_{1i}\,c\alpha_1\,s\theta_1 + F_{1j}\,c\alpha_1\,c\theta_1 + F_{1k}\,s\alpha_1 - F_{2j} \tag{507}$$

$$\begin{aligned}0 = & -T_{1i}\,c\alpha_1\,s\theta_1 + T_{1j}\,c\alpha_1\,c\theta_1 + T_{1k}\,s\alpha_1 - T_{2j} \\ & + a_1\left(F_{1i}\,s\alpha_1\,s\theta_1 + F_{1j}\,s\alpha_1\,c\theta_1 + F_{1k}\,c\alpha_1\right) \\ & - s_1\left(F_{1i}\,c\alpha_1\,c\theta_1 + F_{1j}\,c\alpha_1\,s\theta_1\right)\end{aligned} \tag{508}$$

$$0 = F_{1i}\,s\alpha_1\,s\theta_1 - F_{1j}\,s\alpha_1\,c\theta_1 + F_{1k}\,c\alpha_1 - F_{2k} \tag{509}$$

$$\begin{aligned}0 = & -T_{1i}\,s\alpha_1\,s\theta_1 - T_{1j}\,s\alpha_1\,c\theta_1 + T_{1k}\,c\alpha_1 \\ & + a_1\left(F_{1i}\,c\alpha_1\,s\theta_1 - F_{1j}\,c\alpha_1\,c\theta_1 - F_{1k}\,s\alpha_1\right) \\ & + s_1\left(F_{1i}\,s\alpha_1\,c\theta_1 + F_{1j}\,s\alpha_1\,s\theta_1\right)\end{aligned} \tag{510}$$

$$0 = F_{2i}\,c\theta_2 + F_{2j}\,s\theta_2 - F_{3i} \tag{511}$$

$$0 = T_{2i}\,c\theta_2 + T_{2j}\,s\theta_2 - T_{3i} - s_2\left(F_{2i}\,s\theta_2 - F_{2j}\,c\theta_2\right) \tag{512}$$

$$0 = -F_{2i}\,c\alpha_2\,s\theta_2 + F_{2j}\,c\alpha_2\,c\theta_2 + F_{2k}\,s\alpha_2 - F_{3j} \tag{513}$$

$$\begin{aligned}0 = & -T_{2i}\,c\alpha_2\,s\theta_2 + T_{2j}\,c\alpha_2\,c\theta_2 - T_{3j} \\ & + a_2\left(F_{2i}\,s\alpha_2\,s\theta_2 - F_{2j}\,s\alpha_2\,c\theta_2 + F_{2k}\,c\alpha_2\right) \\ & - s_2\left(F_{2i}\,c\alpha_2\,c\theta_2 + F_{2j}\,c\alpha_2\,s\theta_2\right)\end{aligned} \tag{514}$$

$$0 = F_{2i}\,s\alpha_2\,s\theta_2 - F_{2j}\,s\alpha_2\,c\theta_2 + F_{2k}\,c\alpha_2 - F_{3k} \tag{515}$$

$$\begin{aligned}0 = & \ T_{2i}\,s\alpha_2\,s\theta_2 - T_{2j}\,s\alpha_2\,c\theta_2 \\ & + a_2\left(F_{2i}\,c\alpha_2\,s\theta_2 - F_{2j}\,c\alpha_2\,c\theta_2 - F_{2k}\,s\alpha_2\right) \\ & + s_2\left(F_{2i}\,s\alpha_2\,c\theta_2 + F_{2j}\,s\alpha_2\,s\theta_2\right)\end{aligned} \tag{516}$$

$$0 = F_{3i}\,c\theta_3 + F_{3j}\,s\theta_3 - F_{4i} \tag{517}$$

$$0 = T_{3i}\,c\theta_3 + T_{3j}\,s\theta_3 - T_{4i} - s_3\left(F_{3i}\,s\theta_3 - F_{3j}\,c\theta_3\right) \tag{518}$$

$$0 = -F_{3i}\,c\alpha_3\,s\theta_3 + F_{3j}\,c\alpha_3\,c\theta_3 + F_{3k}\,s\alpha_3 - F_{4j} \tag{519}$$

$$\begin{aligned}0 = & -T_{3i}\,c\alpha_3\,s\theta_3 + T_{3j}\,c\alpha_3\,c\theta_3 - T_{4j} \\ & + a_3\left(F_{3i}\,s\alpha_3\,s\theta_3 - F_{3j}\,s\alpha_3\,c\theta_3 + F_{3k}\,c\alpha_3\right) \\ & - s_3\left(F_{3i}\,c\alpha_3\,c\theta_3 + F_{3j}\,c\alpha_3\,s\theta_3\right)\end{aligned} \tag{520}$$

$$0 = F_{3i}\,s\alpha_3\,s\theta_3 - F_{3j}\,s\alpha_3\,c\theta_3 + F_{3k}\,c\alpha_3 - F_{4k} \tag{521}$$

$$\begin{aligned} 0 = {} & T_{3i}\, s\alpha_3\, s\theta_3 - T_{3j}\, s\alpha_3\, c\theta_3 - T_{4k} \\ & + a_3\left(F_{3i}\, c\alpha_3\, s\theta_3 - F_{3j}\, c\alpha_3\, c\theta_3 - F_{3k}\, s\alpha_3\right) \\ & + s_3\left(F_{3i}\, s\alpha_3\, c\theta_3 + F_{3j}\, s\alpha_3\, s\theta_3\right) \end{aligned} \tag{522}$$

While there are 18 equations, considering input-joint torque T_{1k} and force F_{1k} as the input variables leaves 20 unknowns. The usual procedure to resolve the indeterminacy is to consider the intermediate-joint axial forces F_{2k} and F_{3k} as given rather than as unknown quantities.

The 18 equations can be organized into two subsets of 9 equations each, one subset containing equations with only force variables while the other contains both the force and torque variables. The force subset can be solved first and the results plugged into the other subset which can then be solved for the torque variables.

The force subset can be written in matrix form as

$$\begin{bmatrix} c\theta_1 & s\theta_1 & -1 & 0 & 0 & 0 & 0 & 0 & 0 \\ -c\alpha_1 s\theta_1 & c\alpha_1 c\theta_1 & 0 & -1 & 0 & 0 & 0 & 0 & 0 \\ s\alpha_1 s\theta_1 & -s\alpha_1 c\theta_1 & 0 & 0 & 0 & 0 & 0 & 0 & 0 \\ 0 & 0 & c\theta_2 & s\theta_2 & -1 & 0 & 0 & 0 & 0 \\ 0 & 0 & -c\alpha_2 s\theta_2 & c\alpha_2 c\theta_2 & 0 & -1 & 0 & 0 & 0 \\ 0 & 0 & s\alpha_2 s\theta_2 & -s\alpha_2 c\theta_2 & 0 & 0 & 0 & 0 & 0 \\ 0 & 0 & 0 & 0 & c\theta_3 & s\theta_3 & -1 & 0 & 0 \\ 0 & 0 & 0 & 0 & -c\alpha_3 s\theta_3 & c\alpha_3 c\theta_3 & 0 & -1 & 0 \\ 0 & 0 & 0 & 0 & s\alpha_3 s\theta_3 & -s\alpha_3 c\theta_3 & 0 & 0 & -1 \end{bmatrix} \begin{Bmatrix} F_{1i} \\ F_{1j} \\ F_{2i} \\ F_{2j} \\ F_{3i} \\ F_{3j} \\ F_{4i} \\ F_{4j} \\ F_{4k} \end{Bmatrix} = \begin{Bmatrix} 0 \\ -F_{1k}\, s\alpha_1 \\ -F_{1k}\, c\alpha_1 + F_{2k} \\ 0 \\ -F_{2k}\, s\alpha_2 \\ -F_{2k}\, c\alpha_2 + F_{3k} \\ 0 \\ -F_{3k}\, s\alpha_3 \\ -F_{3k}\, c\alpha_3 \end{Bmatrix} \tag{523}$$

or

$$[C]\{F\} = \{K_F\} \tag{524}$$

The solutions are

$$F_{2i} = Q_{F5} \tag{525}$$

$$F_{3j} = Q_{F2} - Q_{F1} F_{2i} \tag{526}$$

$$F_{1i} = K_{F1} - F_{2i}\, c\theta_1 - F_{3j}\, s\theta_1 \tag{527}$$

$$F_{1j} = K_{F2} + F_{2i}\, c\alpha_1\, s\theta_1 - F_{3j}\, c\alpha_1\, c\theta_1 \tag{528}$$

$$F_{2j} = K_{F4} - F_{1i}\, c\theta_1 - F_{1j}\, s\theta_2 \tag{529}$$

$$F_{3i} = K_{F5} + F_{1i}\, c\alpha_2\, s\theta_2 - F_{1j}\, c\alpha_2\, c\theta_2 \tag{530}$$

$$F_{4i} = K_{F7} - F_{2j}\, c\theta_3 - F_{3i}\, s\theta_3 \tag{531}$$

$$F_{4j} = K_{F8} + F_{2j}\, c\alpha_3\, s\theta_3 - F_{3i}\, c\alpha_3\, c\theta_3 \tag{532}$$

$$F_{4k} = K_{F9} - F_{2j}\,\mathrm{s}\alpha_3\,\mathrm{s}\theta_3 + F_{3i}\,\mathrm{s}\alpha_3\,\mathrm{c}\theta_3 \tag{533}$$

where

$$Q_{F1} = -\frac{\mathrm{c}\alpha_1\,\mathrm{s}\theta_1\,\mathrm{c}\theta_2 + \mathrm{c}\theta_1\,\mathrm{s}\theta_2}{\mathrm{c}\alpha_1\,\mathrm{c}\theta_1\,\mathrm{c}\theta_2 + \mathrm{s}\theta_1\,\mathrm{s}\theta_2} \tag{534}$$

$$Q_{F2} = \frac{K_{F6} - \mathrm{s}\alpha_2\left(K_{F2}\,\mathrm{c}\theta_2 - K_{F1}\,\mathrm{s}\theta_2\right)}{\mathrm{s}\alpha_2\left(\mathrm{c}\alpha_1\,\mathrm{c}\theta_1\,\mathrm{c}\theta_2 - \mathrm{s}\theta_1\,\mathrm{s}\theta_2\right)} \tag{535}$$

$$Q_{F3} = \mathrm{s}\alpha_1\,\mathrm{s}\theta_1 + Q_{F1}\,\mathrm{s}\alpha_1\,\mathrm{c}\theta_1 \tag{536}$$

$$Q_{F4} = K_{F3} + Q_{F2}\,\mathrm{s}\alpha_1\,\mathrm{c}\theta_1 \tag{537}$$

$$Q_{F5} = \frac{Q_{F4}}{Q_{F3}} \tag{538}$$

where factors K_{Fi} are the elements of vector $\{\mathrm{K}_F\}$.

The torque subset written in matrix form is

$$\begin{bmatrix} \mathrm{c}\theta_1 & \mathrm{s}\theta_1 & -1 & 0 & 0 & 0 & 0 & 0 & 0 \\ -\mathrm{c}\alpha_1\,\mathrm{s}\theta_1 & \mathrm{c}\alpha_1\,\mathrm{c}\theta_1 & 0 & -1 & 0 & 0 & 0 & 0 & 0 \\ \mathrm{s}\alpha_1\,\mathrm{s}\theta_1 & -\mathrm{s}\alpha_1\,\mathrm{c}\theta_1 & 0 & 0 & 0 & 0 & 0 & 0 & 0 \\ 0 & 0 & \mathrm{c}\theta_2 & \mathrm{s}\theta_2 & -1 & 0 & 0 & 0 & 0 \\ 0 & 0 & -\mathrm{c}\alpha_2\,\mathrm{s}\theta_2 & \mathrm{c}\alpha_2\,\mathrm{c}\theta_2 & 0 & -1 & 0 & 0 & 0 \\ 0 & 0 & \mathrm{s}\alpha_2\,\mathrm{s}\theta_2 & -\mathrm{s}\alpha_2\,\mathrm{c}\theta_2 & 0 & 0 & 0 & 0 & 0 \\ 0 & 0 & 0 & 0 & \mathrm{c}\theta_3 & \mathrm{s}\theta_3 & -1 & 0 & 0 \\ 0 & 0 & 0 & 0 & -\mathrm{c}\alpha_3\,\mathrm{s}\theta_3 & \mathrm{c}\alpha_3\,\mathrm{c}\theta_3 & 0 & -1 & 0 \\ 0 & 0 & 0 & 0 & \mathrm{s}\alpha_3\,\mathrm{s}\theta_3 & -\mathrm{s}\alpha_3\,\mathrm{c}\theta_3 & 0 & 0 & -1 \end{bmatrix} \begin{Bmatrix} T_{1i} \\ T_{1j} \\ T_{2i} \\ T_{2j} \\ T_{3i} \\ T_{3j} \\ T_{4i} \\ T_{4j} \\ T_{4k} \end{Bmatrix}$$

$$= \begin{Bmatrix} s_1\left(F_{1i}\,\mathrm{s}\theta_3 - F_{1j}\,\mathrm{c}\theta_3\right) \\ -T_{1k}\,\mathrm{s}\alpha_1 - a_1\left(F_{1i}\,\mathrm{s}\alpha_1\,\mathrm{s}\theta_1 - F_{1j}\,\mathrm{s}\alpha_1\,\mathrm{c}\theta_1 + F_{1k}\,\mathrm{c}\alpha_1\right) + s_1\left(F_{1i}\,\mathrm{c}\alpha_1\,\mathrm{c}\theta_1 + F_{1j}\,\mathrm{c}\alpha_1\,\mathrm{s}\theta_1\right) \\ -T_{1k}\,\mathrm{c}\alpha_1 - a_1\left(F_{1i}\,\mathrm{c}\alpha_1\,\mathrm{s}\theta_1 - F_{1j}\,\mathrm{c}\alpha_1\,\mathrm{c}\theta_1 - F_{1k}\,\mathrm{s}\alpha_1\right) - s_1\left(F_{1i}\,\mathrm{s}\alpha_1\,\mathrm{c}\theta_1 + F_{1j}\,\mathrm{s}\alpha_1\,\mathrm{s}\theta_1\right) \\ -s_2\left(F_{2i}\,\mathrm{s}\theta_2 + F_{2j}\,\mathrm{c}\theta_2\right) \\ -a_2\left(F_{2i}\,\mathrm{s}\alpha_2\,\mathrm{s}\theta_2 - F_{2j}\,\mathrm{s}\alpha_2\,\mathrm{c}\theta_2\right) - s_2\left(F_{2i}\,\mathrm{c}\alpha_2\,\mathrm{c}\theta_2 + F_{2j}\,\mathrm{c}\alpha_2\,\mathrm{s}\theta_2\right) \\ -a_2\left(F_{2i}\,\mathrm{c}\alpha_2\,\mathrm{s}\theta_2 - F_{2j}\,\mathrm{c}\alpha_2\,\mathrm{c}\theta_2\right) - s_2\left(F_{2i}\,\mathrm{s}\alpha_2\,\mathrm{c}\theta_2 + F_{2j}\,\mathrm{s}\alpha_2\,\mathrm{s}\theta_2\right) \\ -s_3\left(F_{3i}\,\mathrm{s}\theta_3 + F_{3j}\,\mathrm{c}\theta_3\right) \\ -a_3\left(F_{3i}\,\mathrm{s}\alpha_3\,\mathrm{s}\theta_3 - F_{3j}\,\mathrm{s}\alpha_3\,\mathrm{c}\theta_3 + F_{3k}\,\mathrm{c}\alpha_3\right) + s_3\left(F_{3i}\,\mathrm{c}\alpha_3\,\mathrm{c}\theta_3 + F_{3j}\,\mathrm{c}\alpha_3\,\mathrm{s}\theta_3\right) \\ -a_3\left(F_{3i}\,\mathrm{c}\alpha_3\,\mathrm{s}\theta_3 - F_{3j}\,\mathrm{c}\alpha_3\,\mathrm{c}\theta_3 + F_{3k}\,\mathrm{s}\alpha_3\right) + s_3\left(F_{3i}\,\mathrm{s}\alpha_3\,\mathrm{c}\theta_3 + F_{3j}\,\mathrm{s}\alpha_3\,\mathrm{s}\theta_3\right) \end{Bmatrix} \tag{539}$$

or

$$[\mathrm{C}]\{\mathrm{T}\} = \{\mathrm{K}_T\} \tag{540}$$

It is observed that the coefficient matrices of both the force subset, eq.(523), and the torque subset, eq.(539), are identical. Consequently the solution of the torque subset will have the same form as the solutions of the force subset,

$$T_{2i} = Q_{T5} \tag{541}$$

$$T_{3j} = Q_{T2} - Q_{T1} T_{2i} \tag{542}$$

$$T_{1i} = K_{T1} - T_{2i}\, c\theta_1 - T_{3j}\, s\theta_1 \tag{543}$$

$$T_{1j} = K_{T2} + T_{2i}\, c\alpha_1\, s\theta_1 - T_{3j}\, c\alpha_1\, c\theta_1 \tag{544}$$

$$T_{2j} = K_{T4} - T_{1i}\, c\theta_1 - T_{1j}\, s\theta_2 \tag{545}$$

$$T_{3i} = K_{T5} + T_{1i}\, c\alpha_2\, s\theta_2 - T_{1j}\, c\alpha_2\, c\theta_2 \tag{546}$$

$$T_{4i} = K_{T7} - T_{2j}\, c\theta_3 - T_{3i}\, s\theta_3 \tag{547}$$

$$T_{4j} = K_{T8} + T_{2j}\, c\alpha_3\, s\theta_3 - T_{3i}\, c\alpha_3\, c\theta_3 \tag{548}$$

$$T_{4k} = K_{T9} - T_{2j}\, s\alpha_3\, s\theta_3 + T_{3i}\, s\alpha_3\, c\theta_3 \tag{549}$$

where

$$Q_{T1} = -\frac{c\alpha_1\, s\theta_1\, c\theta_2 + c\theta_1\, s\theta_2}{c\alpha_1\, c\theta_1\, c\theta_2 + s\theta_1\, s\theta_2} \tag{550}$$

$$Q_{T2} = \frac{K_{T6} - s\alpha_2\left(K_{T2}\, c\theta_2 - K_{T1}\, s\theta_2\right)}{s\alpha_2\left(c\alpha_1\, c\theta_1\, c\theta_2 - s\theta_1\, s\theta_2\right)} \tag{551}$$

$$Q_{T3} = s\alpha_1\, s\theta_1 + Q_{T1}\, s\alpha_1\, c\theta_1 \tag{552}$$

$$Q_{T4} = K_{T3} + Q_{T2}\, s\alpha_1\, c\theta_1 \tag{553}$$

$$Q_{T5} = \frac{Q_{T4}}{Q_{T3}} \tag{554}$$

where factors K_{Ti} are elements of vector $\{K_T\}$.

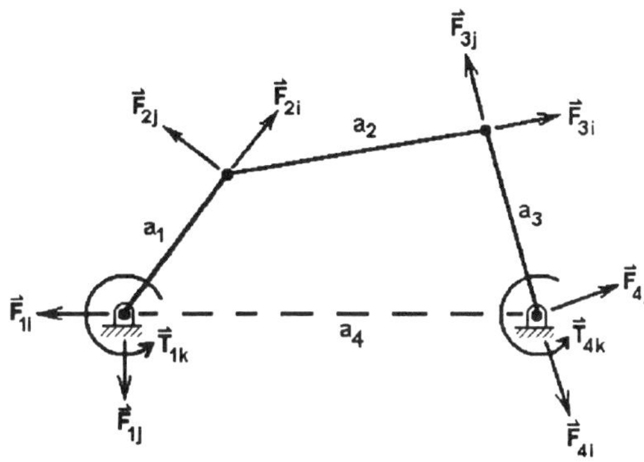

Figure 53 Static forces and torques of the planar four-bar linkage.

Planar Four-Bar Linkage

For a link with two revolute joints and no twist, the equation of static equilibrium

$$0 = {}^{n+1}_{n}\hat{\mathrm{M}}\,{}^{n}\hat{\mathrm{F}}_{n} - {}^{n+1}\hat{\mathrm{F}}_{n+1} \tag{555}$$

becomes

$$\begin{Bmatrix}0\\0\\0\end{Bmatrix} = \begin{bmatrix} c\theta_n & s\theta_n & 0 \\ -s\theta_n & c\theta_n & \varepsilon a_n \\ \varepsilon a_n s\theta_n & -\varepsilon a_n c\theta_n & 1 \end{bmatrix} \begin{Bmatrix} {}^{n}F_{n,i} + \varepsilon\,{}^{n}T_{n,i} \\ {}^{n}F_{n,j} + \varepsilon\,{}^{n}T_{n,j} \\ {}^{n}F_{n,k} + \varepsilon\,{}^{n}T_{n,k} \end{Bmatrix} - \begin{Bmatrix} {}^{n+1}F_{n+1,i} + \varepsilon\,{}^{n+1}T_{n+1,i} \\ {}^{n+1}F_{n+1,j} + \varepsilon\,{}^{n+1}T_{n+1,j} \\ {}^{n+1}F_{n+1,k} + \varepsilon\,{}^{n+1}T_{n+1,k} \end{Bmatrix} \tag{556}$$

$$\begin{Bmatrix}0\\0\\0\end{Bmatrix} = \begin{Bmatrix} {}^{n}F_{n,i}\,c\theta_n + {}^{n}F_{n,j}\,s\theta_n - {}^{n+1}F_{n+1,i} + \varepsilon\left({}^{n}T_{n,i}\,c\theta_n + {}^{n}T_{n,j}\,s\theta_n - {}^{n+1}T_{n+1,i}\right) \\ -{}^{n}F_{n,i}\,s\theta_n + {}^{n}F_{n,j}\,c\theta_n - {}^{n+1}F_{n+1,j} + \varepsilon\left(-{}^{n}T_{n,i}\,s\theta_n + {}^{n}T_{n,j}\,c\theta_n - {}^{n+1}T_{n+1,j} + a_n\,{}^{n}F_{n,k}\right) \\ {}^{n}F_{n,k} - {}^{n+1}F_{n+1,k} + \varepsilon\left({}^{n}T_{n,k} - {}^{n+1}T_{n+1,k} + a_n\,{}^{n}F_{n,i}\,s\theta_n - a_n\,{}^{n}F_{n,j}\,c\theta_n\right) \end{Bmatrix} \tag{557}$$

The equations of equilibrium of the frictionless planar four-bar linkage shown in Figure 53 are developed from eq.(557).

Under static conditions and in the absence of friction, an intermediate revolute joint cannot support a torque about its axis. Therefore

$${}^{2}T_{2k} = 0 \tag{558}$$

$${}^{3}T_{3k} = 0 \tag{559}$$

In typical practice, there are no forces applied along the joint axes so that

$${}^{1}F_{1k} = 0 \tag{560}$$

$$ {}^{2}F_{2k} = 0 \tag{561} $$

$$ {}^{3}F_{3k} = 0 \tag{562} $$

$$ {}^{4}F_{4k} = 0 \tag{563} $$

Furthermore, there are no torques transmitted in the plane of the mechanism,

$$ {}^{1}T_{1i} = 0 \tag{564} $$

$$ {}^{1}T_{1j} = 0 \tag{565} $$

$$ {}^{2}T_{2i} = 0 \tag{566} $$

$$ {}^{2}T_{2j} = 0 \tag{567} $$

$$ {}^{3}T_{3i} = 0 \tag{568} $$

$$ {}^{3}T_{3j} = 0 \tag{569} $$

$$ {}^{4}T_{4i} = 0 \tag{570} $$

$$ {}^{4}T_{4j} = 0 \tag{571} $$

Substituting eqs.(558-571) into eq.(557) and evaluating for each of the three moving links produces a set of nine non-trivial scalar equations.

$$ 0 = {}^{1}F_{1i}\, c\theta_1 + {}^{1}F_{1j}\, s\theta_1 - {}^{2}F_{2i} \tag{572} $$

$$ 0 = -{}^{1}F_{1i}\, s\theta_1 + {}^{1}F_{1j}\, c\theta_1 - {}^{2}F_{2j} \tag{573} $$

$$ 0 = {}^{1}T_{1k} + {}^{1}F_{1i} a_1\, s\theta_1 - {}^{1}F_{1j} a_1\, c\theta_1 \tag{574} $$

$$ 0 = {}^{2}F_{2i}\, c\theta_2 + {}^{2}F_{2j}\, s\theta_2 - {}^{3}F_{3i} \tag{575} $$

$$ 0 = -{}^{2}F_{2i}\, s\theta_2 + {}^{2}F_{2j}\, c\theta_2 - {}^{3}F_{3j} \tag{576} $$

$$ 0 = {}^{2}F_{2i} a_2\, s\theta_2 - {}^{2}F_{2j} a_2\, c\theta_2 \tag{577} $$

$$ 0 = {}^{3}F_{3i}\, c\theta_3 + {}^{3}F_{3j}\, s\theta_3 - {}^{4}F_{4i} \tag{578} $$

$$ 0 = -{}^{3}F_{3i}\, s\theta_3 + {}^{3}F_{3j}\, c\theta_3 - {}^{4}F_{4j} \tag{579} $$

$$ 0 = -{}^{4}T_{4k} + {}^{3}F_{3i} a_3\, s\theta_3 - {}^{3}F_{3j} a_3\, c\theta_3 \tag{580} $$

This can be put in matrix form as

$$\begin{bmatrix} c\theta_1 & s\theta_1 & -1 & 0 & 0 & 0 & 0 & 0 & 0 \\ -s\theta_1 & c\theta_1 & 0 & -1 & 0 & 0 & 0 & 0 & 0 \\ a_1 s\theta_1 & -a_1 c\theta_1 & 0 & 0 & 0 & 0 & 0 & 0 & 0 \\ 0 & 0 & c\theta_2 & s\theta_2 & -1 & 0 & 0 & 0 & 0 \\ 0 & 0 & -s\theta_2 & c\theta_2 & 0 & -1 & 0 & 0 & 0 \\ 0 & 0 & a_2 s\theta_2 & -a_2 c\theta_2 & 0 & 0 & 0 & 0 & 0 \\ 0 & 0 & 0 & 0 & c\theta_3 & s\theta_3 & -1 & 0 & 0 \\ 0 & 0 & 0 & 0 & -s\theta_3 & c\theta_3 & 0 & -1 & 0 \\ 0 & 0 & 0 & 0 & a_3 s\theta_3 & -a_3 c\theta_3 & 0 & 0 & -1 \end{bmatrix} \begin{Bmatrix} {}^1F_{1i} \\ {}^1F_{1j} \\ {}^2F_{2i} \\ {}^2F_{2j} \\ {}^3F_{3i} \\ {}^3F_{3j} \\ {}^4F_{4i} \\ {}^4F_{4j} \\ {}^4T_{4k} \end{Bmatrix} = \begin{Bmatrix} 0 \\ 0 \\ -{}^1T_{1k} \\ 0 \\ 0 \\ 0 \\ 0 \\ 0 \\ 0 \end{Bmatrix} \quad (581)$$

The solutions of this set of equations are

$$\begin{Bmatrix} {}^1F_{1i} \\ {}^1F_{1j} \\ {}^2F_{2i} \\ {}^2F_{2j} \\ {}^3F_{3i} \\ {}^3F_{3j} \\ {}^4F_{4i} \\ {}^4F_{4j} \\ {}^4T_{4k} \end{Bmatrix} = {}^1T_{1k} \begin{Bmatrix} \dfrac{\dfrac{c\theta_1}{\tan\theta_2} - s\theta_1}{a_1} \\ \dfrac{c\theta_1 + \dfrac{s\theta_1}{\tan\theta_2}}{a_1} \\ \dfrac{\cot\theta_2}{a_1} \\ \dfrac{1}{a_1} \\ \dfrac{1}{a_1 s\theta_2} \\ \dfrac{c\theta_3}{a_1 s\theta_2} \\ -\dfrac{s\theta_3}{a_1 s\theta_2} \\ \dfrac{a_3 s\theta_3}{a_1 s\theta_2} \end{Bmatrix} \quad (582)$$

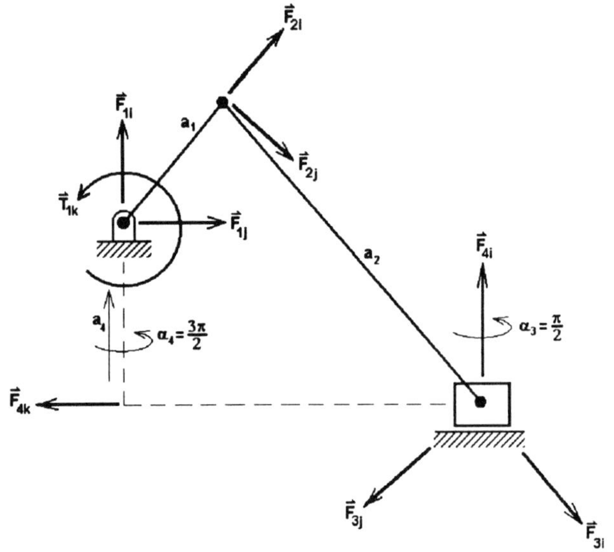

Figure 54 Static forces and torques of the planar slider-crank mechanism.

Planar Slider-Crank Mechanism

The joint-link transformation matrices for the planar slider-crank mechanism illustrated in Figure 54 are

$$ {}^{1}_{2}\hat{M} = \begin{bmatrix} c\theta_1 & -s\theta_1 & \varepsilon a_1 s\theta_1 \\ s\theta_1 & c\theta_1 & -\varepsilon a_1 c\theta_1 \\ 0 & \varepsilon a_1 & 1 \end{bmatrix} \tag{583} $$

$$ {}^{2}_{3}\hat{M} = \begin{bmatrix} c\theta_2 & -s\theta_2 & \varepsilon a_2 s\theta_2 \\ s\theta_2 & c\theta_2 & -\varepsilon a_2 c\theta_2 \\ 0 & \varepsilon a_2 & 1 \end{bmatrix} \tag{584} $$

$$ {}^{3}_{4}\hat{M} = \begin{bmatrix} c\theta_3 & 0 & s\theta_3 \\ s\theta_3 & 0 & -c\theta_3 \\ 0 & 1 & 0 \end{bmatrix} \tag{585} $$

$$ {}^{4}_{1}\hat{M} = \begin{bmatrix} 1 & 0 & -\varepsilon s_4 \\ \varepsilon s_4 & \varepsilon a_4 & 1 \\ 0 & -1 & \varepsilon a_4 \end{bmatrix} \tag{586} $$

The input torque and bearing reactions are contained in the dual force

$$ {}^1\hat{\mathrm{F}}_1 = \begin{Bmatrix} {}^1F_{1i} \\ {}^1F_{1j} \\ \varepsilon {}^1T_{1k} \end{Bmatrix} \tag{587} $$

Because intermediate revolute joints do not support a torque along their axes, the intermediate reactions are

$$ {}^2\hat{\mathrm{F}}_2 = \begin{Bmatrix} {}^2F_{2i} \\ {}^2F_{2j} \\ 0 \end{Bmatrix} \tag{588} $$

$$ {}^3\hat{\mathrm{F}}_3 = \begin{Bmatrix} {}^3F_{3i} \\ {}^3F_{3j} \\ 0 \end{Bmatrix} \tag{589} $$

The output dual force is

$$ {}^4\hat{\mathrm{F}}_4 = \begin{Bmatrix} {}^4F_{4i} \\ 0 \\ {}^4F_{4k} \end{Bmatrix} \tag{590} $$

Now, for a link with two joints, the equation of static equilibrium is

$$ 0 = {}^{n+1}_{\ \ n}\hat{\mathrm{M}}\ {}^n\hat{\mathrm{F}}_n - {}^{n+1}\hat{\mathrm{F}}_{n+1} \tag{591} $$

For the crank, link 1, we substitute eqs.(583), (587) and (588) into eq.(591)

$$ 0 = {}^2_1\hat{\mathrm{M}}\ {}^1\hat{\mathrm{F}}_1 + {}^2\hat{\mathrm{F}}_2 = \begin{bmatrix} c\theta_1 & s\theta_1 & 0 \\ -s\theta_1 & c\theta_1 & \varepsilon a_1 \\ \varepsilon a_1 s\theta_1 & -\varepsilon a_1 c\theta_1 & 1 \end{bmatrix} \begin{Bmatrix} {}^1F_{1i} \\ {}^1F_{1j} \\ \varepsilon {}^1T_{1k} \end{Bmatrix} - \begin{Bmatrix} {}^2F_{2i} \\ {}^2F_{2j} \\ 0 \end{Bmatrix} \tag{592} $$

$$ \begin{Bmatrix} 0 \\ 0 \\ 0 \end{Bmatrix} = \begin{Bmatrix} {}^1F_{1i}\, c\theta_1 + {}^1F_{1j}\, s\theta_1 - {}^2F_{2i} \\ -{}^1F_{1i}\, s\theta_1 + {}^1F_{1j}\, c\theta_1 - {}^2F_{2j} \\ \varepsilon\left(a_1\, {}^1F_{1i}\, s\theta_1 - a_1\, {}^1F_{1j}\, c\theta_1 + {}^1T_{1k}\right) \end{Bmatrix} \tag{593} $$

For the connecting rod, link 2, we substitute eqs.(584), (588) and (589) into eq.(591)

$$0={}_2^3\hat{\mathrm{M}}\,{}^2\hat{\mathrm{F}}_2+{}^3\hat{\mathrm{F}}_3=\begin{bmatrix} c\theta_2 & s\theta_2 & 0 \\ -s\theta_2 & c\theta_2 & \varepsilon a_2 \\ \varepsilon a_2\, s\theta_2 & -\varepsilon a_2\, c\theta_2 & 1 \end{bmatrix}\begin{Bmatrix} {}^2F_{2i} \\ {}^2F_{2j} \\ 0 \end{Bmatrix}-\begin{Bmatrix} {}^3F_{3i} \\ {}^3F_{3j} \\ 0 \end{Bmatrix} \tag{594}$$

$$\begin{Bmatrix} 0 \\ 0 \\ 0 \end{Bmatrix}=\begin{Bmatrix} {}^2F_{2i}\, c\theta_2+{}^2F_{2j}\, s\theta_2-{}^3F_{3i} \\ -{}^2F_{2i}\, s\theta_2+{}^2F_{2j}\, c\theta_2-{}^3F_{3j} \\ \varepsilon\left(a_2\, {}^2F_{2i}\, s\theta_2-a_2\, {}^2F_{2j}\, c\theta_2\right) \end{Bmatrix} \tag{595}$$

For the slider, link 3, we substitute eqs.(585), (589) and (590) into eq.(591)

$$0={}_3^4\hat{\mathrm{M}}\,{}^3\hat{\mathrm{F}}_3+{}^4\hat{\mathrm{F}}_4=\begin{bmatrix} c\theta_3 & s\theta_3 & 0 \\ 0 & 0 & 1 \\ s\theta_3 & -c\theta_3 & 0 \end{bmatrix}\begin{Bmatrix} {}^3F_{3i} \\ {}^3F_{3j} \\ 0 \end{Bmatrix}-\begin{Bmatrix} {}^4F_{4i} \\ 0 \\ {}^4F_{4k} \end{Bmatrix} \tag{596}$$

$$\begin{Bmatrix} 0 \\ 0 \\ 0 \end{Bmatrix}=\begin{Bmatrix} {}^3F_{3i}\, c\theta_3+{}^3F_{3j}\, s\theta_3-{}^4F_{4i} \\ 0 \\ {}^3F_{3i}\, s\theta_3-{}^3F_{3j}\, c\theta_3-{}^4F_{4k} \end{Bmatrix} \tag{597}$$

Eqs.(593), (595) and (597) constitute a set of 8 scalar equations in crank length a_1, connecting rod length a_2 and input torque ${}^1T_{1k}$, which can be solved for the unknowns ${}^1F_{1i}$, ${}^1F_{1j}$, ${}^2F_{2i}$, ${}^2F_{2j}$, ${}^3F_{3i}$, ${}^3F_{3j}$, ${}^4F_{4i}$, ${}^4F_{4j}$.

Written in matrix form, these equations form the set

$$\begin{bmatrix} c\theta_1 & s\theta_1 & -1 & 0 & 0 & 0 & 0 & 0 \\ -s\theta_1 & c\theta_1 & 0 & -1 & 0 & 0 & 0 & 0 \\ a_1\, s\theta_1 & a_1\, c\theta_1 & 0 & 0 & 0 & 0 & 0 & 0 \\ 0 & 0 & c\theta_2 & s\theta_2 & -1 & 0 & 0 & 0 \\ 0 & 0 & -s\theta_2 & c\theta_2 & 0 & -1 & 0 & 0 \\ 0 & 0 & a_2\, s\theta_2 & a_2\, c\theta_2 & 0 & 0 & 0 & 0 \\ 0 & 0 & 0 & 0 & c\theta_3 & s\theta_3 & -1 & 0 \\ 0 & 0 & 0 & 0 & s\theta_3 & -c\theta_3 & 0 & -1 \end{bmatrix}\begin{Bmatrix} {}^1F_{1i} \\ {}^1F_{1j} \\ {}^2F_{2i} \\ {}^2F_{2j} \\ {}^3F_{3i} \\ {}^3F_{3j} \\ {}^4F_{4i} \\ {}^4F_{4k} \end{Bmatrix}=\begin{Bmatrix} 0 \\ 0 \\ -{}^1T_{1k} \\ 0 \\ 0 \\ 0 \\ 0 \\ 0 \end{Bmatrix} \tag{598}$$

The solutions of this equation set are

$${}^1F_{1i}=\left[\frac{c(\theta_1+2\theta_2)+c\theta_1}{a_1\, s(2\theta_1)}\right]{}^1T_{1k} \tag{599}$$

$${}^1F_{1j}=\left[\frac{s(\theta_1+2\theta_2)+s\theta_1}{a_1\, s(2\theta_1)}\right]{}^1T_{1k} \tag{600}$$

$$ {}^{2}F_{2i} = \frac{{}^{1}T_{1k}}{a_1 \tan\theta_2} \tag{601} $$

$$ {}^{2}F_{2j} = \frac{{}^{1}T_{1k}}{a_1} \tag{602} $$

$$ {}^{3}F_{3i} = \frac{{}^{1}T_{1k}}{a_1\, \mathrm{s}\theta_2} \tag{603} $$

$$ {}^{3}F_{3j} = 0 \tag{604} $$

$$ {}^{4}F_{4i} = \frac{\mathrm{c}\theta_3}{a_1\, \mathrm{s}\theta_2}\, {}^{1}T_{1k} \tag{605} $$

$$ {}^{4}F_{4k} = \frac{\mathrm{s}\theta_3}{a_1\, \mathrm{s}\theta_2}\, {}^{1}T_{1k} \tag{606} $$

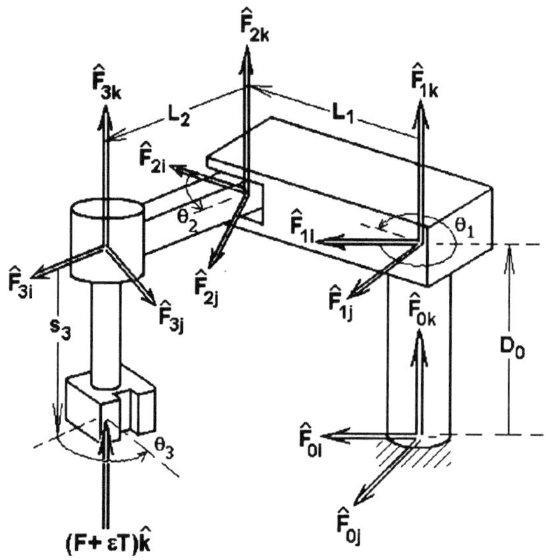

Figure 55 A robot manipulator subjected to a load $(F+\varepsilon T)\hat{\mathrm{k}}$.

Robotic Manipulator

Consider the evaluation of the joint and base reactions for the robot shown in Figure 55 when a load is applied on the hand such that

$$ {}^{4}\hat{\mathrm{F}}_{4} = \begin{Bmatrix} 0 \\ 0 \\ F+\varepsilon\, T \end{Bmatrix} \tag{607} $$

Now, the equation of static equilibrium for each link is

$$ 0 = {}^{n+1}_{n}\hat{\mathrm{M}}\ {}^{n}\hat{\mathrm{F}}_{n} - {}^{n+1}\hat{\mathrm{F}}_{n+1} \tag{608} $$

For an open chain where the load is applied at the free end, it is easier to multiply eq.(608) by matrix ${}^{\;\mathrm{n}}_{\mathrm{n+1}}\hat{\mathrm{M}}$ and rearrange as

$$ {}^{n}\hat{\mathrm{F}}_{n} = {}^{\;n}_{n+1}\hat{\mathrm{M}}\,{}^{n+1}\hat{\mathrm{F}}_{n+1} \tag{609} $$

Starting at the hand,

$$ {}^{3}\hat{F}_{3} = {}^{3}_{4}M^{4}\hat{F}_{4} = \begin{bmatrix} c\theta_3 - \varepsilon s_3 s\theta_3 & -s\theta_3 - \varepsilon s_3 c\theta_3 & 0 \\ s\theta_3 + \varepsilon s_3 c\theta_3 & c\theta_3 - \varepsilon s_3 s\theta_3 & 0 \\ 0 & 0 & 1 \end{bmatrix} \begin{Bmatrix} 0 \\ 0 \\ F + \varepsilon T \end{Bmatrix} = \begin{Bmatrix} 0 \\ 0 \\ F + \varepsilon T \end{Bmatrix} \quad (610) $$

Moving to the elbow,

$$ {}^{2}\hat{F}_{2} = {}^{2}_{3}\hat{M}^{3}F_{3} = \begin{bmatrix} c\theta_2 & -s\theta_2 & \varepsilon L_2 s\theta_2 \\ s\theta_2 & c\theta_2 & -\varepsilon L_2 c\theta_2 \\ 0 & \varepsilon L_2 & 1 \end{bmatrix} \begin{Bmatrix} 0 \\ 0 \\ F + \varepsilon T \end{Bmatrix} = \begin{Bmatrix} \varepsilon F L_2 s\theta_2 \\ -\varepsilon F L_2 c\theta_2 \\ F + \varepsilon T \end{Bmatrix} \quad (611) $$

Progressing to the shoulder,

$$ {}^{1}\hat{F}_{1} = -{}^{1}_{2}M^{2}\hat{F}_{2} = \begin{bmatrix} c\theta_1 & -s\theta_1 & \varepsilon L_1 s\theta_1 \\ s\theta_1 & c\theta_1 & -\varepsilon L_1 c\theta_1 \\ 0 & \varepsilon L_1 & 1 \end{bmatrix} \begin{Bmatrix} \varepsilon F L_2 s\theta_2 \\ -\varepsilon F L_2 c\theta_2 \\ F + \varepsilon T \end{Bmatrix} \quad (612) $$

$$ {}^{1}\hat{F}_{1} = \begin{Bmatrix} -\varepsilon F\left[L_1 s\theta_1 + L_2 s(\theta_1 + \theta_2)\right] \\ \varepsilon F\left[L_1 c\theta_1 + L_2 c(\theta_1 + \theta_2)\right] \\ F + \varepsilon T \end{Bmatrix} \quad (613) $$

At the base,

$$ {}^{0}\hat{F}_{0} = {}^{0}_{1}M^{1}\hat{F}_{1} = \begin{bmatrix} 1 & -\varepsilon D_1 & 0 \\ \varepsilon D_1 & 1 & 0 \\ 0 & 0 & 1 \end{bmatrix} \begin{Bmatrix} -\varepsilon\left[L_1 s\theta_1 + L_2 s(\theta_1 + \theta_2)\right] \\ \varepsilon\left[L_1 c\theta_1 + L_2 c(\theta_1 + \theta_2)\right] \\ F + \varepsilon T \end{Bmatrix} \quad (614) $$

$$ {}^{0}\hat{F}_{0} = \begin{Bmatrix} -\varepsilon F\left[L_1 s\theta_1 + L_2 s(\theta_1 + \theta_2)\right] \\ \varepsilon F\left[L_1 c\theta_1 + L_2 c(\theta_1 + \theta_2)\right] \\ F + \varepsilon T \end{Bmatrix} \quad (615) $$

This result represents the force exerted by the robot on the ground.

Exercises - Static Analysis

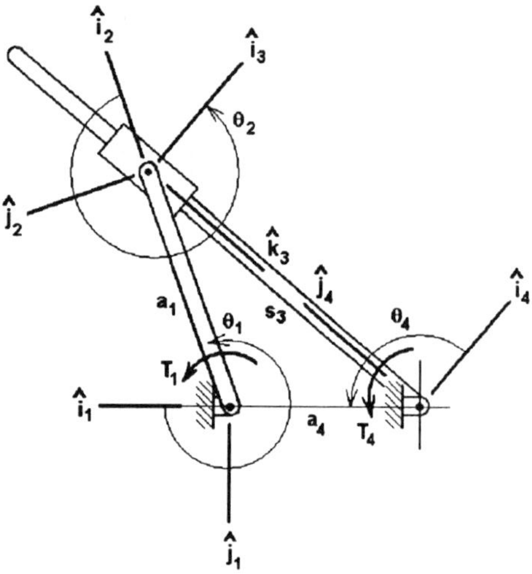

Figure 56 Swinging-block mechanism.

1. Obtain the relationship between output torque T_4 and input torque T_1 of the swinging-block mechanism shown in Figure 56.

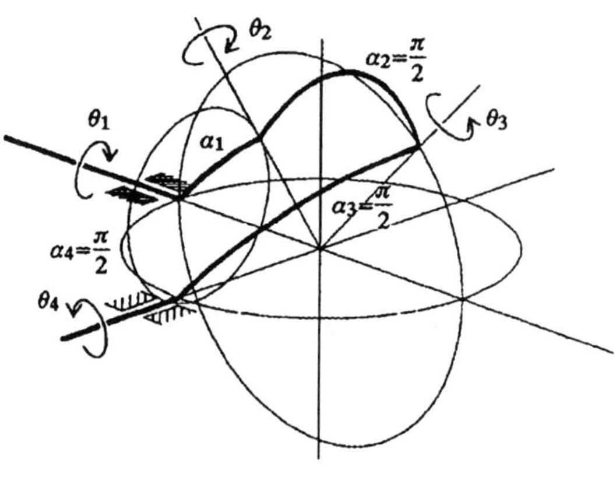

Figure 57 Wobble-plate mechanism.

2. Perform a static analysis for the wobble-plate mechanism, a spherical four-bar mechanism where the input crank has some twist α_1 and the other links have twists $\alpha_2=\alpha_3=\alpha_4=\pi/2$. Such a mechanism is shown in Figure 57.

Acceleration

If a point P on the screw axis $\vec{u}$ of the motion of a rigid body has a dual velocity

$$\hat{V}^P = (\Omega + \varepsilon V)\vec{u} \qquad (616)$$

then application of the equation of transformation gives the velocity of that rigid body at a point Q as

$$\hat{V}^Q = \Omega\vec{u} + \varepsilon\left[V\vec{u} + \Omega(\vec{u} \times \vec{\rho})\right] \qquad (617)$$

where vector $\vec{\rho}$ extends from point P on the screw axis to point Q. Acceleration is defined as the time derivative of velocity, so that the acceleration of the rigid body at point Q is

$$\hat{A}^Q = \frac{d}{dt}\hat{V}^Q = \frac{d}{dt}\left\{\Omega\vec{u} + \varepsilon\left[V\vec{u} + \Omega(\vec{u} \times \vec{\rho})\right]\right\} \qquad (618)$$

$$\hat{A}^Q = \frac{d}{dt}\Omega\vec{u} + \varepsilon\left[\frac{d}{dt}V\vec{u} + \left(\frac{d}{dt}\Omega\right)(\vec{u} \times \vec{\rho}) + \Omega\left(\vec{u} \times \frac{d}{dt}\vec{\rho}\right)\right] \qquad (619)$$

Let symbol $\alpha = \dot{\Omega}$ represent the angular acceleration of the body about the screw axis and symbol $a = \dot{V}$ represent the linear acceleration of point P on the body along the screw axis. Note that the screw axis $\vec{u}$ is a vector which is at least instantaneously constant, while vector $\vec{\rho}$ locating point Q in the body relative to point P on the screw axis will move along with the body in which it is fixed, so that

$$\frac{d}{dt}\vec{u} = 0 \qquad (620)$$

and

$$\frac{d}{dt}\vec{\rho} = \dot{\vec{\rho}} + \Omega(\vec{u} \times \vec{\rho}) = \Omega\vec{u} \times \vec{\rho} \qquad (621)$$

where velocity $\dot{\vec{\rho}} = 0$ on account of the rigid-body condition.

Thus, eq.(619) can be put in the form

$$\hat{A}^Q = \alpha\vec{u} + \varepsilon\left[a\vec{u} + \alpha(\vec{u} \times \vec{\rho}) - \Omega^2\vec{\rho}\right] \qquad (622)$$

Each term in this equation has physical significance. Term $\alpha\bar{u}$ represents the angular acceleration of the body about the screw axis. The dual terms are linear accelerations. Term $a\bar{u}$ represents the acceleration of the sliding along the screw axis, term $\alpha(\bar{u}\times\bar{\rho})$ corresponds to the acceleration tangential to the path of point Q and term $-\Omega^2\bar{\rho}$ indicates the acceleration in direction normal to the path of point Q.

Because of the normal component, dual acceleration is not a dual vector. A comparison of the expression for dual acceleration with that for dual velocity will show the difference. Because of this extra term, accelerations at different points on a rigid body cannot be obtained by application of the equation of transformation. For this reason, Yang (1966) has characterized dual acceleration as a "pseudo dual vector."

Typically, dynamic analysis is based on accelerations, but because dual acceleration is not a dual vector, it will have little utility for us, and we will instead consider dual momentum in our formulation of dynamic behavior.

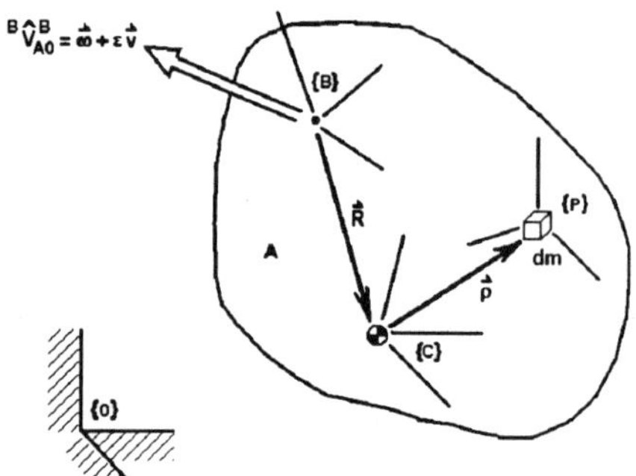

Figure 58 Rigid body A in motion relative to fixed frame $\{0\}$.

Dual Momentum

Consider a rigid body A in motion relative to ground O as shown in Figure 58. This body consists of elements dm at the points P comprising the body such that the total mass is

$$m_A = \int_A dm \tag{623}$$

and the velocity of body A at point B relative to ground O in terms of a frame $\{B\}$ with origin at point B is

$$^{B}\hat{V}^{B}_{A0} = \vec{\omega} + \varepsilon\vec{v} \tag{624}$$

where vector $\vec{v}$ represents the linear velocity and vector $\vec{\omega}$ represents the angular velocity at point *B*, both vectors expressed in terms of frame {B}.

The dual velocity of body *A* relative to ground *0* at a point *P* in terms of a frame {P} oriented the same as frame {B} with origin at that point *P* is obtained by application of the equation of transformation as

$$^{B=P}\hat{V}^{P}_{A0} = \vec{\omega} + \varepsilon\left[\vec{v} - \left(\vec{R} + {}^{B}_{C}T_{P}\vec{\rho}\right) \times \vec{\omega}\right] \tag{625}$$

In this expression, vector $\vec{R}$ extends from point *B* to a reference point *C* and is expressed in terms of the unit vectors of frame {B}. Vector $\vec{\rho}$ runs from point *C* to point *P* and is expressed in the unit vectors of a frame {C} with origin at point *C* whose orientation relative to frame {B} is specified by transformation matrix ${}^{B}_{C}T_{P}$. The negative sign arises from choosing the direction of vectors from the point of known velocity to the point of unknown velocity.

The linear momentum of a particle *P* at point *P* of rigid body *A* relative to the ground *0* with respect to point *B* in terms of frame {B} is defined as

$$^{B}\vec{P}_{P} = \left[\vec{v} - \left(\vec{R} + {}^{B}_{C}T_{P}\vec{\rho}\right) \times \vec{\omega}\right]dm \tag{626}$$

where the particle has an elemental mass *dm*. The angular momentum, or moment of momentum, of a particle with elemental mass *dm* at a point *P* in rigid body *A* in terms of frame {B} is

$$^{B}\vec{H}_{P} = \left(\vec{R} + {}^{B}_{C}T_{P}\vec{\rho}\right) \times {}^{B}\vec{P}_{P} \tag{627}$$

$$^{B}\vec{H}_{P} = \left(\vec{R} + {}^{B}_{C}T_{P}\vec{\rho}\right) \times \left[\vec{v} - \left(\vec{R} + {}^{B}_{C}T_{P}\vec{\rho}\right) \times \omega\right]dm \tag{628}$$

The dual momentum of the particle is

$$^{B}\hat{H}_{P} = {}^{B}\vec{P}_{P} + \varepsilon\,{}^{B}\vec{H}_{P} \tag{629}$$

The dual momentum of a rigid body *A* in frame {B} is the sum of the dual momenta of its constitutive particles

$$^{B}\hat{H}_{A} = \int_{A}\left({}^{B}\vec{P}_{P} + \varepsilon\,{}^{B}\vec{H}_{P}\right) \tag{630}$$

Substituting eqs.(626) and (628) into eq.(630) yields

$$
{}^{B}\hat{\mathrm{H}}_{A}=\int_{A}\left[\vec{\mathrm{v}}-\left(\vec{\mathrm{R}}+{}_{C}^{B}\mathrm{T}_{p}\vec{\rho}\right)\times\vec{\omega}\right]dm+\varepsilon\int_{A}\left(\vec{\mathrm{R}}+{}_{C}^{B}\mathrm{T}_{p}\vec{\rho}\right)\times\left[\vec{\mathrm{v}}-\left(\vec{\mathrm{R}}+{}_{C}^{B}\mathrm{T}_{p}\vec{\rho}\right)\times\omega\right]dm \tag{631}
$$

$$
\begin{aligned}
&=\int_{A}\vec{\mathrm{v}}\,dm-\int_{A}\vec{\mathrm{R}}\times\vec{\omega}\,dm-\int_{A}{}_{C}^{B}\mathrm{T}_{p}\vec{\rho}\times\vec{\omega}\,dm\\
&+\\
&\varepsilon\left\{\begin{array}{c}\displaystyle\int_{A}\vec{\mathrm{R}}\times\vec{\mathrm{v}}\,dm+\int_{A}{}_{C}^{B}\mathrm{T}_{p}\vec{\rho}\times\vec{\mathrm{v}}\,dm\\ \displaystyle-\int_{A}\vec{\mathrm{R}}\times\vec{\mathrm{R}}\times\vec{\omega}\,dm-\int_{A}\vec{\mathrm{R}}\times{}_{C}^{B}\mathrm{T}_{p}\vec{\rho}\times\vec{\omega}\,dm-\int_{A}{}_{C}^{B}\mathrm{T}_{p}\vec{\rho}\times\vec{\mathrm{R}}\times\vec{\omega}\,dm-\int_{A}{}_{C}^{B}\mathrm{T}_{p}\vec{\rho}\times{}_{C}^{B}\mathrm{T}_{p}\vec{\rho}\times\vec{\omega}\,dm\end{array}\right\}
\end{aligned} \tag{632}
$$

$$
\begin{aligned}
&=\int_{A}\vec{\mathrm{v}}\,dm-\int_{A}\vec{\mathrm{R}}\times\vec{\omega}\,dm-\int_{A}{}_{C}^{B}\mathrm{T}_{p}\vec{\rho}\times\vec{\omega}\,dm\\
&+\\
&\varepsilon\left\{\begin{array}{c}\displaystyle\int_{A}\vec{\mathrm{R}}\times\vec{\mathrm{v}}\,dm+\int_{A}{}_{C}^{B}\mathrm{T}_{p}\vec{\rho}\times\vec{\mathrm{v}}\,dm\\ \displaystyle-\int_{A}\vec{\mathrm{R}}\times\vec{\mathrm{R}}\times\vec{\omega}\,dm-\int_{A}\vec{\mathrm{R}}\times{}_{C}^{B}\mathrm{T}_{p}\vec{\rho}\times\vec{\omega}\,dm+\int_{A}\vec{\mathrm{R}}\times{}_{C}^{B}\mathrm{T}_{p}\vec{\rho}\times\vec{\omega}\,dm-\int_{A}{}_{C}^{B}\mathrm{T}_{p}\vec{\rho}\times{}_{C}^{B}\mathrm{T}_{p}\vec{\rho}\times\vec{\omega}\,dm\end{array}\right\}
\end{aligned} \tag{633}
$$

$$
\begin{aligned}
&=\int_{A}\vec{\mathrm{v}}\,dm-\int_{A}\vec{\mathrm{R}}\times\vec{\omega}\,dm-\int_{A}{}_{C}^{B}\mathrm{T}_{p}\vec{\rho}\times\vec{\omega}\,dm\\
&+\\
&\varepsilon\left\{\int_{A}\vec{\mathrm{R}}\times\vec{\mathrm{v}}\,dm+\int_{A}{}_{C}^{B}\mathrm{T}_{p}\vec{\rho}\times\vec{\mathrm{v}}\,dm-\int_{A}\vec{\mathrm{R}}\times\vec{\mathrm{R}}\times\vec{\omega}\,dm-\int_{A}{}_{C}^{B}\mathrm{T}_{p}\vec{\rho}\times{}_{C}^{B}\mathrm{T}_{p}\vec{\rho}\times\vec{\omega}\,dm\right\}
\end{aligned} \tag{634}
$$

We now make the substitutions

$$
\vec{\mathrm{R}}\times=\begin{bmatrix}0 & -R_{z} & R_{y}\\ R_{z} & 0 & -R_{x}\\ -R_{y} & R_{x} & 0\end{bmatrix} \tag{635}
$$

and

$$
{}_{C}^{B}\mathrm{T}_{p}\vec{\rho}\times={}_{C}^{B}\mathrm{T}_{p}\begin{bmatrix}0 & -\rho_{z} & \rho_{y}\\ \rho_{z} & 0 & -\rho_{x}\\ -\rho_{y} & \rho_{x} & 0\end{bmatrix}{}_{C}^{B}\mathrm{T}_{p}^{-1} \tag{636}
$$

and so,

$$
{}^{B}\hat{\mathrm{H}}_{A} = \int_{A}\vec{\mathrm{v}}\,dm - \int_{A}\begin{bmatrix} 0 & -R_z & R_y \\ R_z & 0 & -R_x \\ -R_y & R_x & 0 \end{bmatrix}\vec{\omega}\,dm - \int_{A} {}^{B}_{C}\mathrm{T}_p \begin{bmatrix} 0 & -\rho_z & \rho_y \\ \rho_z & 0 & -\rho_x \\ -\rho_y & \rho_x & 0 \end{bmatrix} {}^{B}_{C}\mathrm{T}_p^{-1}\vec{\omega}\,dm
$$

$$
+\;\varepsilon\left[\begin{array}{c} \int_{A}\begin{bmatrix} 0 & -R_z & R_y \\ R_z & 0 & -R_x \\ -R_y & R_x & 0 \end{bmatrix}\vec{\mathrm{v}}\,dm + \int_{A} {}^{B}_{C}T_p \begin{bmatrix} 0 & -\rho_z & \rho_y \\ \rho_z & 0 & -\rho_x \\ -\rho_y & \rho_x & 0 \end{bmatrix} {}^{B}_{C}\mathrm{T}_p^{-1}\vec{\mathrm{v}}\,dm \\ -\int_{A}\begin{bmatrix} 0 & -R_z & R_y \\ R_z & 0 & -R_x \\ -R_y & R_x & 0 \end{bmatrix}\begin{bmatrix} 0 & -R_z & R_y \\ R_z & 0 & -R_x \\ -R_y & R_x & 0 \end{bmatrix}\vec{\omega}\,dm \\ -\int_{A} {}^{B}_{C}\mathrm{T}_p \begin{bmatrix} 0 & -\rho_z & \rho_y \\ \rho_z & 0 & -\rho_x \\ -\rho_y & \rho_x & 0 \end{bmatrix} {}^{B}_{C}\mathrm{T}_p^{-1}\,{}^{B}_{C}\mathrm{T}_p \begin{bmatrix} 0 & -\rho_z & \rho_y \\ \rho_z & 0 & -\rho_x \\ -\rho_y & \rho_x & 0 \end{bmatrix} {}^{B}_{C}\mathrm{T}_p^{-1}\vec{\omega}\,dm \end{array}\right] \tag{637}
$$

Now, the velocities at point B, $\vec{\mathrm{v}}$ and $\vec{\omega}$, vector $\vec{\mathrm{R}}$ and the rotation operator ${}^{B}_{C}\mathrm{T}_p$ are constant for all points of the rigid body A, so therefore they can be taken outside of the integrals.

$$
{}^{B}\hat{\mathrm{H}}_{A} = m_A\vec{\mathrm{v}} - \left\{ m_A \begin{bmatrix} 0 & -R_z & R_y \\ R_z & 0 & -R_x \\ -R_y & R_x & 0 \end{bmatrix} \right\}\vec{\omega} + {}^{B}_{C}\mathrm{T}_p \left\{ \int_{A} \begin{bmatrix} 0 & -\rho_z & \rho_y \\ \rho_z & 0 & -\rho_x \\ -\rho_y & \rho_x & 0 \end{bmatrix} dm \right\} {}^{B}_{C}\mathrm{T}_p^{-1}\vec{\omega}
$$

$$
+\;\varepsilon\left[\begin{array}{c} m_A\left\{\begin{bmatrix} 0 & -R_z & R_y \\ R_z & 0 & -R_x \\ -R_y & R_x & 0 \end{bmatrix}\right\}\vec{\mathrm{v}} + {}^{B}_{C}\mathrm{T}_p \left\{ \int_{A} \begin{bmatrix} 0 & -\rho_z & \rho_y \\ \rho_z & 0 & -\rho_x \\ -\rho_y & \rho_x & 0 \end{bmatrix} dm \right\} {}^{B}_{C}\mathrm{T}_p^{-1}\vec{\mathrm{v}} \\ -\left\{ m_A \begin{bmatrix} 0 & -R_z & R_y \\ R_z & 0 & -R_x \\ -R_y & R_x & 0 \end{bmatrix}\begin{bmatrix} 0 & -R_z & R_y \\ R_z & 0 & -R_x \\ -R_y & R_x & 0 \end{bmatrix} \right\}\vec{\omega} \\ -{}^{B}_{C}\mathrm{T}_p \left\{ \int_{A} \begin{bmatrix} 0 & -\rho_z & \rho_y \\ \rho_z & 0 & -\rho_x \\ -\rho_y & \rho_x & 0 \end{bmatrix}\begin{bmatrix} 0 & -\rho_z & \rho_y \\ \rho_z & 0 & -\rho_x \\ -\rho_y & \rho_x & 0 \end{bmatrix} dm \right\} {}^{B}_{C}\mathrm{T}_p^{-1}\vec{\omega} \end{array}\right] \tag{638}
$$

$$
\begin{aligned}
&{}^{B}\hat{\mathrm{H}}_{A} \\
&= \\
&m_A\vec{\mathrm{v}}-\left\{m_A\begin{bmatrix}0 & -R_z & R_y\\ R_z & 0 & -R_x\\ -R_y & R_x & 0\end{bmatrix}\right\}\vec{\omega}+{}^{B}_{C}\mathrm{T}_{p}\left\{\int_{a}\begin{bmatrix}0 & -\rho_z & \rho_y\\ \rho_z & 0 & -\rho_x\\ -\rho_y & \rho_x & 0\end{bmatrix}dm\right\}{}^{B}_{C}\mathrm{T}_{p}^{-1}\,\vec{\omega} \\
&+ \\
&\varepsilon\left[\begin{gathered}
m_A\left\{\begin{bmatrix}0 & -R_z & R_y\\ R_z & 0 & -R_x\\ -R_y & R_x & 0\end{bmatrix}+{}^{B}_{C}\mathrm{T}_{p}\int_{A}\begin{bmatrix}0 & -\rho_z & \rho_y\\ \rho_z & 0 & -\rho_x\\ -\rho_y & \rho_x & 0\end{bmatrix}dm\,{}^{B}_{C}\mathrm{T}_{p}^{-1}\right\}\vec{\mathrm{v}} \\
-\left\{\begin{gathered}
m_A\begin{bmatrix}0 & -R_z & R_y\\ R_z & 0 & -R_x\\ -R_y & R_x & 0\end{bmatrix}\begin{bmatrix}0 & -R_z & R_y\\ R_z & 0 & -R_x\\ -R_y & R_x & 0\end{bmatrix} \\
+{}^{B}_{C}\mathrm{T}_{p}\int_{A}\begin{bmatrix}0 & -\rho_z & \rho_y\\ \rho_z & 0 & -\rho_x\\ -\rho_y & \rho_x & 0\end{bmatrix}\begin{bmatrix}0 & -\rho_z & \rho_y\\ \rho_z & 0 & -\rho_x\\ -\rho_y & \rho_x & 0\end{bmatrix}dm\ {}^{B}_{C}\mathrm{T}_{p}^{-1}
\end{gathered}\right\}\vec{\omega}
\end{gathered}\right]
\end{aligned}
\tag{639}
$$

$$
\begin{aligned}
&{}^{B}\hat{\mathrm{H}}_{A} \\
&= \\
&m_A\vec{\mathrm{v}}-\left\{m_A\begin{bmatrix}0 & -R_z & R_y\\ R_z & 0 & -R_x\\ -R_y & R_x & 0\end{bmatrix}\right\}\vec{\omega}+{}^{B}_{C}\mathrm{T}_{p}\left\{\int_{A}\begin{bmatrix}0 & -\rho_z & \rho_y\\ \rho_z & 0 & -\rho_x\\ -\rho_y & \rho_x & 0\end{bmatrix}dm\right\}{}^{B}_{C}\mathrm{T}_{p}^{-1}\,\vec{\omega} \\
&+ \\
&\varepsilon\left[\begin{gathered}
m_A\left\{\begin{bmatrix}0 & -R_z & R_y\\ R_z & 0 & -R_x\\ -R_y & R_x & 0\end{bmatrix}+{}^{B}_{C}\mathrm{T}_{p}\int_{A}\begin{bmatrix}0 & -\rho_z & \rho_y\\ \rho_z & 0 & -\rho_x\\ -\rho_y & \rho_x & 0\end{bmatrix}dm\,{}^{B}_{C}\mathrm{T}_{p}^{-1}\right\}\vec{\mathrm{v}} \\
-\left\{\begin{gathered}
m_A\begin{bmatrix}-R_y^2-R_z^2 & R_xR_y & R_xR_z\\ R_xR_y & -R_x^2-R_z^2 & R_yR_z\\ R_xR_z & R_yR_z & -R_x^2-R_y^2\end{bmatrix} \\
+{}^{B}_{C}\mathrm{T}_{p}\int_{A}\begin{bmatrix}-\rho_y^2-\rho_z^2 & \rho_x\rho_y & \rho_x\rho_z\\ \rho_x\rho_y & -\rho_x^2-\rho_z^2 & \rho_y\rho_z\\ \rho_x\rho_z & \rho_y\rho_z & -\rho_x^2-\rho_y^2\end{bmatrix}dm\ {}^{B}_{C}\mathrm{T}_{p}^{-1}
\end{gathered}\right\}\vec{\omega}
\end{gathered}\right]
\end{aligned}
\tag{640}
$$

$$
\begin{aligned}
&{}^{B}\hat{\mathrm{H}}_{A} \\
&= \\
&m_A\vec{\mathrm{v}} - \left\{ m_A \begin{bmatrix} 0 & -R_z & R_y \\ R_z & 0 & -R_x \\ -R_y & R_x & 0 \end{bmatrix} + {}^{B}_{C}\mathrm{T}_p \begin{bmatrix} 0 & -\int_A \rho_z dm & \int_A \rho_y dm \\ \int_A \rho_z dm & 0 & -\int_A \rho_x dm \\ -\int_A \rho_y dm & \int_A \rho_x dm & 0 \end{bmatrix} {}^{B}_{C}\mathrm{T}_p^{-1} \right\} \vec{\omega} \\
&+ \\
&\varepsilon \left[\begin{matrix} m_A \left\{ \begin{bmatrix} 0 & -R_z & R_y \\ R_z & 0 & -R_x \\ -R_y & R_x & 0 \end{bmatrix} + {}^{B}_{C}\mathrm{T}_p \begin{bmatrix} 0 & -\int_A \rho_z dm & \int_A \rho_y dm \\ \int_A \rho_z dm & 0 & -\int_A \rho_x dm \\ -\int_A \rho_y dm & \int_A \rho_x dm & 0 \end{bmatrix} {}^{B}_{C}\mathrm{T}_p^{-1} \right\} \vec{\mathrm{v}} \\
+ \left\{ \begin{matrix} m_A \begin{bmatrix} -R_y^2 - R_z^2 & R_x R_y & R_x R_z \\ R_x R_y & -R_x^2 - R_z^2 & R_y R_z \\ R_x R_z & R_y R_z & -R_x^2 - R_y^2 \end{bmatrix} \\ + {}^{B}_{C}\mathrm{T}_p \begin{bmatrix} -\int_A \left(\rho_y^2 + \rho_z^2\right) dm & \int_A \rho_x \rho_y dm & \int_A \rho_x \rho_z dm \\ \int_A \rho_x \rho_y dm & -\int_A \left(\rho_x^2 + \rho_z^2\right) dm & \int_A \rho_y \rho_z dm \\ \int_A \rho_x \rho_z dm & \int_A \rho_y \rho_z dm & -\int_A \left(\rho_x^2 + \rho_y^2\right) dm \end{bmatrix} {}^{B}_{C}\mathrm{T}_p^{-1} \end{matrix} \right\} \vec{\omega} \end{matrix} \right]
\end{aligned} \tag{641}
$$

The first moments of inertia of body A with respect to point C, in terms of the unit vectors of frame {C} are defined as

$$ {}^{C}S_{Ax}^{C} = \int_A \rho_x dm \tag{642} $$

$$ {}^{C}S_{Ay}^{C} = \int_A \rho_y dm \tag{643} $$

$$ {}^{C}S_{Az}^{C} = \int_A \rho_z dm \tag{644} $$

which can be combined into a matrix

$$ \left[{}^{C}S_{A}^{C}\right] = \begin{bmatrix} 0 & -{}^{C}S_{Az}^{C} & {}^{C}S_{Ay}^{C} \\ {}^{C}S_{Az}^{C} & 0 & -{}^{C}S_{Ax}^{C} \\ -{}^{C}S_{Ay}^{C} & {}^{C}S_{Ax}^{C} & 0 \end{bmatrix} \tag{645} $$

The second moments of inertia of body A with respect to point C in terms of the unit vectors of frame {C} are defined as

$$ {}^{C}J^{C}_{Axx} = \int_A \left(\rho_y^2 + \rho_z^2\right) dm \tag{646} $$

$$ {}^{C}J^{C}_{Ayy} = \int_A \left(\rho_x^2 + \rho_z^2\right) dm \tag{647} $$

$$ {}^{C}J^{C}_{Azz} = \int_A \left(\rho_x^2 + \rho_y^2\right) dm \tag{648} $$

The products of inertia of body A with respect to point C in terms of the unit vectors of frame {C} are defined as

$$ {}^{C}J^{C}_{Axy} = \int_A \rho_x \rho_y dm \tag{649} $$

$$ {}^{C}J^{C}_{Ayz} = \int_A \rho_y \rho_z dm \tag{650} $$

$$ {}^{C}J^{C}_{Axz} = \int_A \rho_x \rho_z dm \tag{651} $$

The second moments of inertia and the products of inertia are combined into an inertia matrix (actually an inertia tensor) defined as

$$ \left[{}^{C}J^{C}_{A}\right] = \begin{bmatrix} {}^{C}J^{C}_{Axx} & -{}^{C}J^{C}_{Axy} & -{}^{C}J^{C}_{Axz} \\ -{}^{C}J^{C}_{Axy} & {}^{C}J^{C}_{Ayy} & -{}^{C}J^{C}_{Ayz} \\ -{}^{C}J^{C}_{Axz} & -{}^{C}J^{C}_{Ayz} & {}^{C}J^{C}_{Azz} \end{bmatrix} \tag{652} $$

The equation for the dual momentum of body A can now be written as

$$ \begin{aligned} {}^{B}\hat{\mathrm{H}}_{A} &= \\ & m_A\vec{\mathrm{v}} - \left\{ m_A \begin{bmatrix} 0 & -R_z & R_y \\ R_z & 0 & -R_x \\ -R_y & R_x & 0 \end{bmatrix} + {}^{B}_{C}\mathrm{T}_p \begin{bmatrix} 0 & -{}^{C}S^{C}_{Az} & {}^{C}S^{C}_{Ay} \\ {}^{C}S^{C}_{Az} & 0 & -{}^{C}S^{C}_{Ax} \\ -{}^{C}S^{C}_{Ay} & {}^{C}S^{C}_{Ax} & 0 \end{bmatrix} {}^{B}_{C}\mathrm{T}_p^{-1} \right\} \vec{\omega} \\ & + \\ & \varepsilon \begin{bmatrix} m_A \left\{ \begin{bmatrix} 0 & -R_z & R_y \\ R_z & 0 & -R_x \\ -R_y & R_x & 0 \end{bmatrix} + {}^{B}_{C}\mathrm{T}_p \begin{bmatrix} 0 & -{}^{C}S^{C}_{Az} & {}^{C}S^{C}_{Ay} \\ {}^{C}S^{C}_{Az} & 0 & -{}^{C}S^{C}_{Ax} \\ -{}^{C}S^{C}_{Ay} & {}^{C}S^{C}_{Ax} & 0 \end{bmatrix} {}^{B}_{C}\mathrm{T}_p^{-1} \right\} \vec{\mathrm{v}} \\ + \left\{ m_A \begin{bmatrix} R_y^2 + R_z^2 & -R_xR_y & -R_xR_z \\ -R_xR_y & R_x^2 + R_z^2 & -R_yR_z \\ -R_xR_z & -R_yR_z & R_x^2 + R_y^2 \end{bmatrix} + {}^{B}_{C}\mathrm{T}_p \begin{bmatrix} {}^{C}J^{C}_{Axx} & -{}^{C}J^{C}_{Axy} & -{}^{C}J^{C}_{Axz} \\ -{}^{C}J^{C}_{Axy} & {}^{C}J^{C}_{Ayy} & -{}^{C}J^{C}_{Ayz} \\ -{}^{C}J^{C}_{Axz} & -{}^{C}J^{C}_{Ayz} & {}^{C}J^{C}_{Azz} \end{bmatrix} {}^{B}_{C}\mathrm{T}_p^{-1} \right\} \vec{\omega} \end{bmatrix} \end{aligned} \tag{653} $$

Now, the first moments of inertia and second moments of inertia of body A with respect to point C in terms of frame {B} are

$$\left[{}^{B}S_{A}^{C}\right] = {}_{C}^{B}\mathrm{T}_{p}\left[{}^{C}S_{A}^{C}\right]{}_{C}^{B}\mathrm{T}_{p}^{-1} \tag{654}$$

$$\left[{}^{B}J_{A}^{C}\right] = {}_{A}^{B}\mathrm{T}_{p}\left[{}^{C}J_{A}^{C}\right]{}_{C}^{B}\mathrm{T}_{p}^{-1} \tag{655}$$

If the transformation matrix ${}_{C}^{B}\mathrm{T}_{p}$ represents a rotation about the z-axis, then the above equations reduce to the equations given in standard texts for stating first and second moments and products of inertia in a coordinate frame with orientation different than that in which the quantities were given.

Thus the equation for the dual momentum can be rewritten as

$$
\begin{aligned}
&{}^{B}\hat{\mathrm{H}}_{A} \\
&= \\
&m_{A}\vec{\mathrm{v}} - \left\{ m_{A}\begin{bmatrix} 0 & -R_{z} & R_{y} \\ R_{z} & 0 & -R_{x} \\ -R_{y} & R_{x} & 0 \end{bmatrix} + \begin{bmatrix} 0 & -{}^{B}S_{Az}^{C} & {}^{B}S_{Ay}^{C} \\ {}^{B}S_{Az}^{C} & 0 & -{}^{B}S_{Ax}^{C} \\ -{}^{B}S_{Ay}^{C} & {}^{B}S_{Ax}^{C} & 0 \end{bmatrix} \right\}\vec{\omega} \\
&+ \\
&\varepsilon\left[\begin{array}{l} m_{A}\left\{ \begin{bmatrix} 0 & -R_{z} & R_{y} \\ R_{z} & 0 & -R_{x} \\ -R_{y} & R_{x} & 0 \end{bmatrix} + \begin{bmatrix} 0 & -{}^{B}S_{Az}^{C} & {}^{B}S_{Ay}^{C} \\ {}^{B}S_{Az}^{C} & 0 & -{}^{B}S_{Ax}^{C} \\ -{}^{B}S_{Ay}^{C} & {}^{B}S_{Ax}^{C} & 0 \end{bmatrix} \right\}\vec{\mathrm{v}} \\
+\left\{ m_{A}\begin{bmatrix} R_{y}^{2}+R_{z}^{2} & -R_{x}R_{y} & -R_{x}R_{z} \\ -R_{x}R_{y} & R_{x}^{2}+R_{z}^{2} & -R_{y}R_{z} \\ -R_{x}R_{z} & -R_{y}R_{z} & R_{x}^{2}+R_{y}^{2} \end{bmatrix} + \begin{bmatrix} {}^{B}J_{Sxx}^{C} & -{}^{B}J_{Axy}^{C} & -{}^{B}J_{Axz}^{C} \\ -{}^{B}J_{Axy}^{C} & {}^{B}J_{Ayy}^{C} & -{}^{B}J_{Ayz}^{C} \\ -{}^{B}J_{Axz}^{C} & -{}^{B}J_{Ayz}^{C} & {}^{B}J_{Azz}^{C} \end{bmatrix} \right\}\vec{\omega} \end{array}\right]
\end{aligned} \tag{656}
$$

$$
{}^{B}\hat{\mathrm{H}}_{A} = \varepsilon\left[\begin{array}{c} m_A\vec{\mathrm{v}} - \begin{bmatrix} 0 & -m_A R_z - {}^{B}S_{Az}^{C} & m_A R_y + {}^{B}S_{Ay}^{C} \\ m_A R_z + {}^{B}S_{Az}^{C} & 0 & -m_A R_x - {}^{B}S_{Ax}^{C} \\ -m_A R_y - {}^{B}S_{Az}^{C} & m_A R_x + {}^{B}S_{Ax}^{C} & 0 \end{bmatrix}\vec{\omega} \\ + \\ \begin{bmatrix} 0 & -m_A R_z - {}^{B}S_{Az}^{C} & m_A R_y + {}^{B}S_{Ay}^{C} \\ m_A R_z + {}^{B}S_{Az}^{C} & 0 & -m_A R_x - {}^{B}S_{Ax}^{C} \\ -m_A R_y - {}^{B}S_{Az}^{C} & m_A R_x + {}^{B}S_{Ax}^{C} & 0 \end{bmatrix}\vec{\mathrm{v}} \\ + \begin{bmatrix} m_A\left(R_z^2 + R_y^2\right) + {}^{B}J_{Axx}^{C} & -m_A R_x R_y - {}^{B}J_{Axy}^{C} & -m_A R_x R_z - {}^{B}J_{Axz}^{C} \\ -m_A R_x R_y - {}^{B}J_{Axy}^{C} & m_A\left(R_x^2 + R_z^2\right) + {}^{B}J_{Ayy}^{C} & -m_A R_y R_z - {}^{B}J_{Ayz}^{C} \\ -m_A R_x R_z - {}^{B}J_{Axz}^{C} & -m_A R_y R_z - {}^{B}J_{Ayz}^{C} & m_A\left(R_x^2 + R_y^2\right) + {}^{B}J_{Azz}^{C} \end{bmatrix}\vec{\omega} \end{array}\right] \tag{657}
$$

Now, the first and second moments of inertia and products of inertia of body A with respect to point B in terms of the unit vectors of frame {B} are

$$ {}^{B}S_{Ax}^{B} = m_A R_x + {}^{B}S_{Ax}^{C} \tag{658} $$

$$ {}^{B}S_{Ay}^{B} = m_A R_y + {}^{B}S_{Ay}^{C} \tag{659} $$

$$ {}^{B}S_{Az}^{B} = m_A R_z + {}^{B}S_{Az}^{C} \tag{660} $$

$$ {}^{B}J_{Axx}^{B} = m_A\left(R_y^2 + R_z^2\right) + {}^{B}J_{Axx}^{C} \tag{661} $$

$$ {}^{B}J_{Ayy}^{B} = m_A\left(R_x^2 + R_z^2\right) + {}^{B}J_{Ayy}^{C} \tag{662} $$

$$ {}^{B}J_{Azz}^{B} = m_A\left(R_x^2 + R_y^2\right) + {}^{B}J_{Azz}^{C} \tag{663} $$

$$ {}^{B}J_{Axy}^{B} = m_A R_x R_y + {}^{B}J_{Axy}^{C} \tag{664} $$

$$ {}^{B}J_{Ayz}^{B} = m_A R_y R_z + {}^{B}J_{Ayz}^{C} \tag{665} $$

$$ {}^{B}J_{Axz}^{B} = m_A R_x R_z + {}^{B}J_{Axz}^{C} \tag{666} $$

These expressions allow first moments, second moments and products of inertia specified in a certain frame to be stated in terms of another frame oriented the same as the first frame but whose origin is displaced relative to the first frame and they constitute the parallel-axis theorem discussed in standard texts.

Combining first moments of inertia into a matrix,

$$\left[{}^{B}S_{A}^{B}\right]=\begin{bmatrix} 0 & -{}^{B}S_{Az}^{B} & {}^{B}S_{Ay}^{B} \\ {}^{B}S_{Az}^{B} & 0 & -{}^{B}S_{Ax}^{B} \\ -{}^{B}S_{Ay}^{B} & {}^{B}S_{Ax}^{B} & 0 \end{bmatrix} \tag{667}$$

and combining the second moments of inertia and the products of inertia into a matrix,

$$\left[{}^{B}J_{A}^{B}\right]=\begin{bmatrix} {}^{B}J_{Axx}^{B} & -{}^{B}J_{Axy}^{B} & -{}^{B}J_{Axz}^{B} \\ -{}^{B}J_{Axy}^{B} & {}^{B}J_{Ayy}^{B} & -{}^{B}J_{Ayz}^{B} \\ -{}^{B}J_{Axz}^{B} & -{}^{B}J_{Ayz}^{B} & {}^{B}J_{Azz}^{B} \end{bmatrix} \tag{668}$$

Substituting these expressions into the equation for the dual momentum of a rigid body,

$${}^{B}\hat{\mathrm{H}}_{A}=m_{A}\vec{\mathrm{v}}-\left[{}^{B}S_{A}^{B}\right]\vec{\omega}+\varepsilon\left(\left[{}^{B}S_{A}^{B}\right]\vec{\mathrm{v}}+\left[{}^{B}J_{A}^{B}\right]\vec{\omega}\right) \tag{669}$$

or

$${}^{B}\hat{\mathrm{H}}_{A}=m_{A}{}^{B}\vec{\mathrm{V}}_{A0d}^{B}-\left[{}^{B}S_{A}^{B}\right]{}^{B}\vec{\mathrm{V}}_{A0p}^{B}+\varepsilon\left(\left[{}^{B}S_{A}^{B}\right]{}^{B}\vec{\mathrm{V}}_{A0d}^{B}+\left[{}^{B}J_{A}^{B}\right]{}^{B}\vec{\mathrm{V}}_{A0p}^{B}\right) \tag{670}$$

where velocities ${}^{B}\vec{\mathrm{V}}_{A0p}^{B}=\vec{\omega}$ and ${}^{B}\vec{\mathrm{V}}_{A0d}^{B}=\vec{\mathrm{v}}$ are the primary and dual components of absolute dual velocity ${}^{B}\hat{\mathrm{V}}_{A0}^{B}$. It is observed that all quantities are expressed in the same frame {B} fixed on the distal end of link A.

Center of Mass and Radii of Gyration

The concepts of center of mass and radii of gyration can often be used to simplify the inertia matrices.

The center of mass of a body is a point selected such that the second terms of the first moments of inertia will vanish. If the center of mass of a link A connecting joints A and B is located in a frame {B} with origin at joint B by a vector

$${}^{B}\vec{\mathrm{G}}_{A}=\begin{Bmatrix} g_{Ax} \\ g_{Ay} \\ g_{Az} \end{Bmatrix} \tag{671}$$

then the first moment of mass matrix of link A in terms of frame {B} is

$$\left[{}^{B}S_{A}^{B}\right]=m_{A}\begin{bmatrix}0 & -g_{Az} & g_{Ay}\\ g_{Az} & 0 & -g_{Ax}\\ -g_{Ay} & g_{Ax} & 0\end{bmatrix} \tag{672}$$

When the center of mass is used as the reference point, the second and third terms in the second moments of inertia will vanish, and the expressions take on the appearance of the parallel axis theorem. Furthermore, the reference frame {B} can be oriented so that the products of inertia vanish, where the frame assumes the orientation of the principal axes. In that case,

$$\left[{}^{B}J_{A}^{B}\right]=m_{a}\begin{bmatrix}k_{Ax}^{2} & 0 & 0\\ 0 & k_{Ay}^{2} & 0\\ 0 & 0 & k_{Az}^{2}\end{bmatrix} \tag{673}$$

where the diagonal terms contain the radii of gyration k_{Ax}, k_{Ay} and k_{Az}.

Inertial Dual Force

Now, the inertia force of a body in motion, ${}^{B}\hat{\mathrm{f}}_{A}$, is obtained by taking the time derivative of its momentum.

$${}^{B}\hat{\mathrm{f}}_{A}=\frac{d}{dt}\left({}^{B}\hat{\mathrm{H}}_{A}\right) \tag{674}$$

$$={}^{B}\dot{\hat{\mathrm{H}}}_{A}+{}^{B}\hat{\mathrm{V}}_{A0}^{B}\times{}^{B}\hat{\mathrm{H}}_{A} \tag{675}$$

$$=\begin{Bmatrix}{}^{B}\dot{\hat{H}}_{Ai}\\ {}^{B}\dot{\hat{H}}_{Aj}\\ {}^{B}\dot{\hat{H}}_{Ak}\end{Bmatrix}+\begin{Bmatrix}{}^{B}\hat{V}_{Ai}^{B}\\ {}^{B}\hat{V}_{Aj}^{B}\\ {}^{B}\hat{V}_{Ak}^{B}\end{Bmatrix}\times\begin{Bmatrix}{}^{B}\hat{H}_{Ai}\\ {}^{B}\hat{H}_{Aj}\\ {}^{B}\hat{H}_{Ak}\end{Bmatrix} \tag{676}$$

$$=\begin{Bmatrix}{}^{B}\dot{\hat{H}}_{Ai}\\ {}^{B}\dot{\hat{H}}_{Aj}\\ {}^{B}\dot{\hat{H}}_{Ak}\end{Bmatrix}+\begin{bmatrix}0 & -{}^{B}\hat{V}_{Ak}^{B} & {}^{B}\hat{V}_{Aj}^{B}\\ {}^{B}\hat{V}_{Ak}^{B} & 0 & -{}^{B}\hat{V}_{Ai}^{B}\\ -{}^{B}\hat{V}_{Aj}^{B} & {}^{B}\hat{V}_{Ai}^{B} & 0\end{bmatrix}\begin{Bmatrix}{}^{B}\hat{H}_{Ai}\\ {}^{B}\hat{H}_{Aj}\\ {}^{B}\hat{H}_{Ak}\end{Bmatrix} \tag{677}$$

This expands to the dual Euler equations

$$ {}^B\hat{\mathrm{f}}_A = \begin{Bmatrix} {}^B\hat{f}_{Ai} \\ {}^B\hat{f}_{Aj} \\ {}^B\hat{f}_{Ak} \end{Bmatrix} = \begin{Bmatrix} {}^B\dot{\hat{H}}_{Ai} - {}^B\hat{V}^B_{Ak}\,{}^B\hat{H}_{Aj} + {}^B\hat{V}^B_{Aj}\,{}^B\hat{H}_{Ak} \\ {}^B\dot{\hat{H}}_{Aj} - {}^B\hat{V}^B_{Ai}\,{}^B\hat{H}_{Ak} + {}^B\hat{V}^B_{Ak}\,{}^B\hat{H}_{Ai} \\ {}^B\dot{\hat{H}}_{Ak} - {}^B\hat{V}^B_{Aj}\,{}^B\hat{H}_{Ai} + {}^B\hat{V}^B_{Ai}\,{}^B\hat{H}_{Aj} \end{Bmatrix} \tag{678} $$

The dual Euler equations are used to calculate the dual-inertia force ${}^B\hat{\mathrm{f}}_A$ associated with the mass and motion of a rigid body A in terms of coordinate frame {B} fixed on the body.

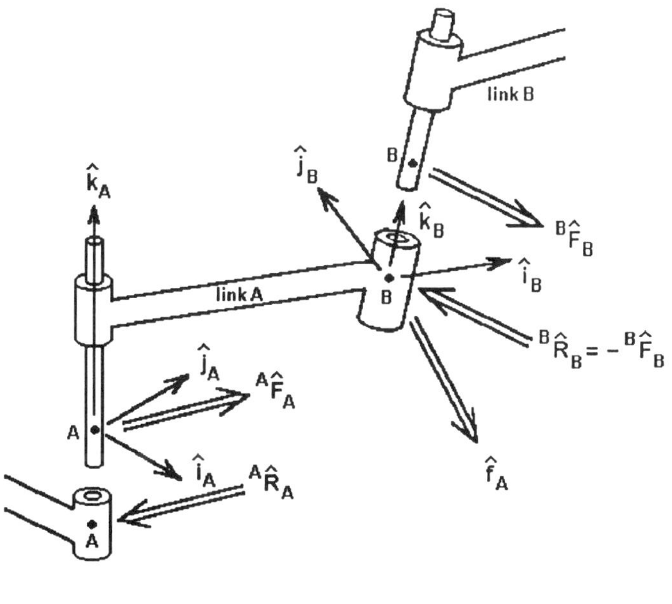

Figure 59 Dual forces acting on link A at its joints A and B.

Dynamic Equilibrium of a Link

D'Alembert's principle can be applied so that this dual-momentum force can be used with the applied dual forces to establish an equation describing the state of dynamic equilibrium of a rigid body. The combination of the inertia dual forces and the applied dual forces together produce equilibrium. Therefore, for a link A connecting joints A and B as shown in Figure 59, the sum of the effect at the distal joint B of dual force ${}^A\hat{\mathrm{F}}_A$ applied at joint A and the reaction $-{}^B\hat{\mathrm{F}}_B$ applied at joint B to the dual force ${}^B\hat{\mathrm{F}}_B$ which link A exerts on link B equals the inertial dual force of link A, all expressed in terms of a frame fixed on the link at the distal joint B,

$$ {}^{B}\hat{\mathrm{f}}_{A} = {}^{B}_{A}\hat{\mathrm{M}}\,{}^{A}\hat{\mathrm{F}}_{A} - {}^{B}\hat{\mathrm{F}}_{B} \tag{679} $$

Dynamic Analysis of Mechanisms

The dual Euler equation can be used to analyze the dynamic behavior of mechanisms using a procedure of the following steps.

Step 1. Obtain the absolute velocity ${}^{B}\hat{\mathrm{V}}^{B}_{A0}$ of each link A connecting joints A and B in terms of the unit vectors of frame {B} with origin at the distal joint.

Step 2. Establish the inertia matrices $\left[{}^{B}S_{A}\right]$ and $\left[{}^{B}J_{A}\right]$ of each moving link A in the mechanism.

Step 3. Obtain the dual momentum ${}^{B}\hat{\mathrm{H}}_{A}$ of each moving link.

Step 4. Differentiate the dual momenta with respect to time to obtain momentum derivatives ${}^{B}\dot{\hat{\mathrm{H}}}_{A}$.

Step 5. Substitute these quantities into the dual Euler equations to obtain the dual-inertial force of each moving link with respect to point B in terms of frame {B}, ${}^{B}\hat{\mathrm{f}}_{A}$.

Step 6. Using the dual-inertial forces and the given input dual forces, write the equation of dynamic equilibrium for each link in terms of the internal and output dual forces by application of D'Alembert's principle.

Step 7. The equations of dynamic equilibrium for the moving links of the mechanism comprise a set of simultaneous equations which can be solved for the internal and output dual forces.

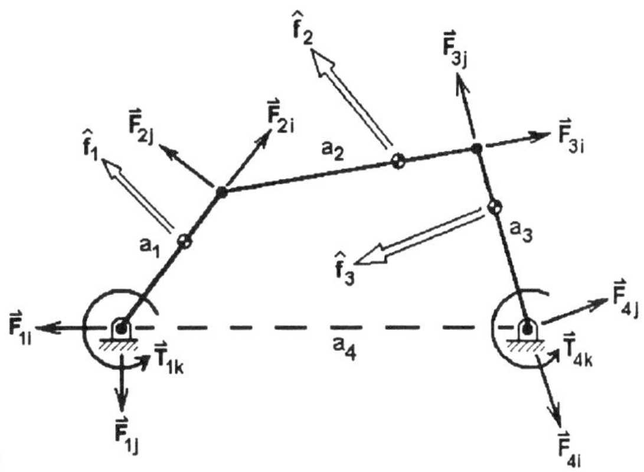

coordi

Figure 60 Dynamic loading of the four-bar mechanism. Although the dual inertia forces are shown acting through a center of mass, in formulating the equations of dynamic equilibrium their effect will be considered at the distal ends of the links.

Planar Four-Bar Mechanism

An analysis of the planar four-bar mechanism shown in Figure 60 is conducted to obtain the absolute velocities of the three moving links in terms of the frames on their distal ends,

$$ {}^{2}\hat{V}_{14}^{2} = \begin{Bmatrix} 0 \\ \varepsilon a_1 \dot{\theta}_1 \\ \dot{\theta}_1 \end{Bmatrix} \tag{680} $$

$$ {}^{3}\hat{V}_{24}^{3} = \begin{Bmatrix} \varepsilon a_1 \dot{\theta}_1 s\theta_2 \\ \varepsilon\left[a_1 \dot{\theta}_1 c\theta_2 + a_2\left(\dot{\theta}_1 + \dot{\theta}_2\right)\right] \\ \dot{\theta}_1 + \dot{\theta}_2 \end{Bmatrix} \tag{681} $$

$$ {}^{4}\hat{V}_{34}^{4} = \begin{Bmatrix} \varepsilon\left[a_1 \dot{\theta}_1 s\theta_{23} + a_2\left(\dot{\theta}_1 + \dot{\theta}_2\right) s\theta_3\right] \\ \varepsilon\left[a_1 \dot{\theta}_1 c\theta_{23} + a_2\left(\dot{\theta}_1 + \dot{\theta}_2\right) c\theta_3 + a_3\left(\dot{\theta}_1 + \dot{\theta}_2 + \dot{\theta}_3\right)\right] \\ \dot{\theta}_1 + \dot{\theta}_2 + \dot{\theta}_3 \end{Bmatrix} \tag{682} $$

If the links are symmetrical about their lengths, then

$$^{2}\vec{G}_1 = \begin{Bmatrix} g_1 \\ 0 \\ 0 \end{Bmatrix}, \quad ^{3}\vec{G}_2 = \begin{Bmatrix} g_2 \\ 0 \\ 0 \end{Bmatrix}, \quad ^{4}\vec{G}_3 = \begin{Bmatrix} g_3 \\ 0 \\ 0 \end{Bmatrix} \tag{683}$$

and so

$$\left[{}^{2}S_1^2\right] = m_1 \begin{bmatrix} 0 & 0 & 0 \\ 0 & 0 & -g_1 \\ 0 & g_1 & 0 \end{bmatrix} \tag{684}$$

$$\left[{}^{3}S_2^3\right] = m_2 \begin{bmatrix} 0 & 0 & 0 \\ 0 & 0 & -g_2 \\ 0 & g_2 & 0 \end{bmatrix} \tag{685}$$

$$\left[{}^{4}S_3^4\right] = m_3 \begin{bmatrix} 0 & 0 & 0 \\ 0 & 0 & -g_3 \\ 0 & g_3 & 0 \end{bmatrix} \tag{686}$$

If the principal axes of each moving link A are oriented the same as the frame {B} fixed on the distal end of its link, then in terms of the radii of gyration k_{Ax}, k_{Ay}, k_{Az}

$$\left[{}^{2}J_1^2\right] = m_1 \begin{bmatrix} k_{1x}^2 & 0 & 0 \\ 0 & k_{1y}^2 & 0 \\ 0 & 0 & k_{1z}^2 \end{bmatrix} \tag{687}$$

$$\left[{}^{3}J_2^3\right] = m_2 \begin{bmatrix} k_{2x}^2 & 0 & 0 \\ 0 & k_{2y}^2 & 0 \\ 0 & 0 & k_{2z}^2 \end{bmatrix} \tag{688}$$

$$\left[{}^{4}J_3^4\right] = m_3 \begin{bmatrix} k_{3x}^2 & 0 & 0 \\ 0 & k_{3y}^2 & 0 \\ 0 & 0 & k_{3z}^2 \end{bmatrix} \tag{689}$$

The dual momentum of each moving link is obtained using the equation

$$^{B}\hat{\text{H}}_A = m_A\, {}^{B}\vec{\text{V}}_{Ad}^{B} - \left[{}^{B}S_A^B\right] {}^{B}\vec{\text{V}}_{Ap}^{B} + \varepsilon\left(\left[{}^{B}S_A^B\right] {}^{B}\vec{\text{V}}_{Ad}^{B} + \left[{}^{B}J_A^B\right] {}^{B}\vec{\text{V}}_{Ap}^{B}\right) \tag{690}$$

and substituting eqs.(680, 684, 687) for the input crank,

$$
\begin{aligned}
&{}^{2}\hat{\mathrm{H}}_{1} \\
&= \\
&m_1\begin{Bmatrix}0\\ a_1\dot{\theta}_1\\ 0\end{Bmatrix}-m_1\begin{bmatrix}0&0&0\\0&0&-g_1\\0&g_1&0\end{bmatrix}\begin{Bmatrix}0\\0\\ \dot{\theta}_1\end{Bmatrix} \\
&+\varepsilon\left(m_1\begin{bmatrix}0&0&0\\0&0&-g_1\\0&g_1&0\end{bmatrix}\begin{Bmatrix}0\\ a_1\dot{\theta}_1\\ 0\end{Bmatrix}+m_1\begin{bmatrix}k_{1x}^2&0&0\\0&k_{1y}^2&0\\0&0&k_{1z}^2\end{bmatrix}\begin{Bmatrix}0\\0\\ \dot{\theta}_1\end{Bmatrix}\right)
\end{aligned}
\tag{691}
$$

$$
{}^{2}\hat{\mathrm{H}}_{1}=m_1\begin{Bmatrix}0\\ \dot{\theta}_1(a_1+g_1)\\ \varepsilon\dot{\theta}_1\left(a_1g_1+k_{1z}^2\right)\end{Bmatrix}
\tag{692}
$$

and eqs.(681, 685, 688) for the coupler,

$$
\begin{aligned}
&{}^{3}\hat{\mathrm{H}}_{2} \\
&= \\
&m_2\begin{Bmatrix}a_1\dot{\theta}_1\,\mathrm{s}\theta_2\\ a_1\dot{\theta}_1\,\mathrm{c}\theta_2+a_2\left(\dot{\theta}_1+\dot{\theta}_2\right)\\ 0\end{Bmatrix}-m_2\begin{bmatrix}0&0&0\\0&0&-g_2\\0&g_2&0\end{bmatrix}\begin{Bmatrix}0\\0\\ \dot{\theta}_1+\dot{\theta}_2\end{Bmatrix} \\
&+\varepsilon\left(m_2\begin{bmatrix}0&0&0\\0&0&-g_2\\0&g_2&0\end{bmatrix}\begin{Bmatrix}a_1\dot{\theta}_1\,\mathrm{s}\theta_2\\ a_1\dot{\theta}_1\,\mathrm{c}\theta_2+a_2\left(\dot{\theta}_1+\dot{\theta}_2\right)\\ 0\end{Bmatrix}+m_2\begin{bmatrix}k_{2x}^2&0&0\\0&k_{2y}^2&0\\0&0&k_{2z}^2\end{bmatrix}\begin{Bmatrix}0\\0\\ \dot{\theta}_1+\dot{\theta}_2\end{Bmatrix}\right)
\end{aligned}
\tag{693}
$$

$$
{}^{3}\hat{\mathrm{H}}_{2}=m_2\begin{Bmatrix}\dot{\theta}_1a_1\,\mathrm{s}\theta_2\\ \dot{\theta}_1a_1\,\mathrm{c}\theta_2+\left(\dot{\theta}_1+\dot{\theta}_2\right)(a_2+g_2)\\ \varepsilon\left[\dot{\theta}_1a_1g_2\,\mathrm{c}\theta_2+\left(\dot{\theta}_2+\dot{\theta}_1\right)\left(a_2g_2+k_{2z}^2\right)\right]\end{Bmatrix}
\tag{694}
$$

and eqs.(682, 686, 689) for the output crank,

$$
\begin{aligned}
&{}^4\hat{\mathrm{H}}_3\\
&\quad :\\
&m_3\begin{Bmatrix} a_1\dot{\theta}_1\,\mathrm{s}\theta_{23}+a_2\left(\dot{\theta}_1+\dot{\theta}_2\right)\mathrm{s}\theta_3 \\ a_1\dot{\theta}_1\,\mathrm{c}\theta_{23}+a_2\left(\dot{\theta}_1+\dot{\theta}_2\right)\mathrm{c}\theta_3+a_3\left(\dot{\theta}_1+\dot{\theta}_2+\dot{\theta}_3\right) \\ 0 \end{Bmatrix}\\
&-m_3\begin{bmatrix} 0 & 0 & 0 \\ 0 & 0 & -g_3 \\ 0 & g_3 & 0 \end{bmatrix}\begin{Bmatrix} 0 \\ 0 \\ \dot{\theta}_1+\dot{\theta}_2+\dot{\theta}_3 \end{Bmatrix}\\
&+\varepsilon\left(\begin{aligned} &m_3\begin{bmatrix} 0 & 0 & 0 \\ 0 & 0 & -g_3 \\ 0 & g_3 & 0 \end{bmatrix}\begin{Bmatrix} a_1\dot{\theta}_1\,\mathrm{s}\theta_{23}+a_2\left(\dot{\theta}_1+\dot{\theta}_2\right)\mathrm{s}\theta_3 \\ a_1\dot{\theta}_1\,\mathrm{c}\theta_{23}+a_2\left(\dot{\theta}_1+\dot{\theta}_2\right)\mathrm{c}\theta_3+a_3\left(\dot{\theta}_1+\dot{\theta}_2+\dot{\theta}_3\right) \\ 0 \end{Bmatrix} \\ &+m_3\begin{bmatrix} k_{3x}^2 & 0 & 0 \\ 0 & k_{3y}^2 & 0 \\ 0 & 0 & k_{3z}^2 \end{bmatrix}\begin{Bmatrix} 0 \\ 0 \\ \dot{\theta}_1+\dot{\theta}_2+\dot{\theta}_3 \end{Bmatrix} \end{aligned}\right)
\end{aligned} \tag{695}
$$

$$
{}^4\hat{\mathrm{H}}_3 = m_3\begin{Bmatrix} \dot{\theta}_1 a_1\,\mathrm{s}\theta_{23}+\left(\dot{\theta}_1+\dot{\theta}_2\right)a_2\,\mathrm{s}\theta_3 \\ \dot{\theta}_1 a_1\,\mathrm{c}\theta_{23}+\left(\dot{\theta}_1+\dot{\theta}_2\right)a_2\,\mathrm{c}\theta_3+\left(\dot{\theta}_1+\dot{\theta}_2+\dot{\theta}_3\right)\left(a_3+g_3\right) \\ \varepsilon\left[\dot{\theta}_1 a_1 g_3\,\mathrm{c}\theta_{23}+\left(\dot{\theta}_1+\dot{\theta}_2\right)a_2 g_3\,\mathrm{c}\theta_3+\left(\dot{\theta}_1+\dot{\theta}_2+\dot{\theta}_3\right)\left(a_3 g_3+k_{3z}^2\right)\right] \end{Bmatrix} \tag{696}
$$

The time derivatives of the dual momenta are obtained by differentiating eqs.(692, 694, 696)

$$
{}^2\dot{\hat{\mathrm{H}}}_1 = m_1\begin{Bmatrix} 0 \\ \ddot{\theta}_1\left(a_1+g_1\right) \\ \varepsilon\ddot{\theta}_1\left(a_1 g_1+k_{1z}^2\right) \end{Bmatrix} \tag{697}
$$

$$
{}^3\dot{\hat{\mathrm{H}}}_2 = m_2\begin{Bmatrix} \ddot{\theta}_1 a_1\,\mathrm{s}\theta_2+\dot{\theta}_1\dot{\theta}_2 a_1\,\mathrm{c}\theta_2 \\ \ddot{\theta}_1 a_1\,\mathrm{c}\theta_2-\dot{\theta}_1\dot{\theta}_2 a_1\,\mathrm{s}\theta_2+\left(\ddot{\theta}_1+\ddot{\theta}_2\right)\left(a_2+g_2\right) \\ \varepsilon\left[\ddot{\theta}_1 a_1 g_2\,\mathrm{c}\theta_2-\dot{\theta}_1\dot{\theta}_2 a_1 g_2\,\mathrm{s}\theta_2+\left(\ddot{\theta}_1+\ddot{\theta}_2\right)\left(a_2 g_2+k_{2z}^2\right)\right] \end{Bmatrix} \tag{698}
$$

$$ {}^{4}\dot{\hat{\mathrm{H}}}_3 = m_3 \begin{Bmatrix} \ddot{\theta}_1 a_1\,\mathrm{s}\theta_{23} + \left(\ddot{\theta}_1+\ddot{\theta}_2\right)a_2\,\mathrm{s}\theta_3 + \dot{\theta}_1\dot{\theta}_2 a_1\,\mathrm{c}\theta_{23} \\ +\dot{\theta}_1\dot{\theta}_3\left(a_1\,\mathrm{c}\theta_{23}+a_2\,\mathrm{c}\theta_3\right)+\dot{\theta}_2\dot{\theta}_3 a_2\,\mathrm{c}\theta_3 \\ \\ \ddot{\theta}_1 a_1\,\mathrm{c}\theta_{23} + \left(\ddot{\theta}_1+\ddot{\theta}_2\right)a_2\,\mathrm{c}\theta_3 + \left(\ddot{\theta}_1+\ddot{\theta}_2+\ddot{\theta}_3\right)\left(a_3+g_3\right) \\ -\dot{\theta}_1\dot{\theta}_2 a_1\,\mathrm{s}\theta_{23} - \dot{\theta}_1\dot{\theta}_3\left(a_1\,\mathrm{s}\theta_{23}+a_2\,\mathrm{s}\theta_3\right)-\dot{\theta}_2\dot{\theta}_3 a_2\,\mathrm{s}\theta_3 \\ \\ \varepsilon\begin{bmatrix} \ddot{\theta}_1 a_1 g_3\,\mathrm{c}\theta_{23} + \left(\ddot{\theta}_1+\ddot{\theta}_2\right)a_2 g_3\,\mathrm{c}\theta_3 + \left(\ddot{\theta}_1+\ddot{\theta}_2+\ddot{\theta}_3\right)\left(a_3 g_3+k_{3z}^2\right) \\ -\dot{\theta}_1\dot{\theta}_2 a_1 g_3\,\mathrm{s}\theta_{23} - \dot{\theta}_1\dot{\theta}_3\left(a_1 g_3\,\mathrm{s}\theta_{23}+a_2 g_3\,\mathrm{s}\theta_3\right)-\dot{\theta}_2\dot{\theta}_3 a_2 g_3\,\mathrm{s}\theta_3 \end{bmatrix} \end{Bmatrix} \tag{699} $$

where values for the joint accelerations $\ddot{\theta}_i$ are obtained from expressions developed by differentiating with respect to time the expressions for the joint speeds $\dot{\theta}_i$ resulting from a velocity analysis.

Next, the dual Euler equation is applied to determine the inertia force associated with each moving link.

For the input crank, eqs.(680, 692, 697) are substituted into the dual Euler equation to obtain

$$ {}^{2}\hat{\mathrm{f}}_1 = m_1 \begin{Bmatrix} -\dot{\theta}_1^2\left(a_1+g_1\right) \\ \ddot{\theta}_1\left(a_1+g_1\right) \\ \varepsilon\ddot{\theta}_1\left(a_1 g_1+k_{1z}^2\right) \end{Bmatrix} \tag{700} $$

for the coupler, eqs.(681, 694, 698) are substituted into the dual Euler equation to obtain

$$ {}^{3}\hat{\mathrm{f}}_2 = m_2 \begin{Bmatrix} \ddot{\theta}_1 a_1\,\mathrm{s}\theta_2 - \dot{\theta}_1^2 a_1\,\mathrm{c}\theta_2 - \left(\dot{\theta}_1+\dot{\theta}_2\right)^2\left(a_2+g_2\right) \\ \\ \ddot{\theta}_1 a_1\,\mathrm{c}\theta_2 + \left(\ddot{\theta}_1+\ddot{\theta}_2\right)\left(a_2+g_2\right)+\dot{\theta}_1^2 a_2\,\mathrm{s}\theta_2 \\ \\ \varepsilon\left[\ddot{\theta}_1 a_1 g_2\,\mathrm{c}\theta_2 + \left(\ddot{\theta}_1+\ddot{\theta}_2\right)\left(a_2 g_2+k_{2z}^2\right)-\dot{\theta}_1\dot{\theta}_2 a_1 g_2\,\mathrm{s}\theta_2\right] \end{Bmatrix} \tag{701} $$

and for the output crank, eqs.(682, 696, 699) are substituted into the dual Euler equation to obtain

$$
{}^{4}\hat{\mathrm{f}}_3 = m_3\left\{\begin{array}{c}
\ddot{\theta}_1 a_1\,\mathrm{s}\theta_{23} + \left(\ddot{\theta}_1 + \ddot{\theta}_2\right) a_2\,\mathrm{s}\theta_3 \\
-\dot{\theta}_1^2 a_1\,\mathrm{c}\theta_{23} - \left(\dot{\theta}_1 + \dot{\theta}_2\right)^2 a_1\,\mathrm{c}\theta_3 - \left(\dot{\theta}_1 + \dot{\theta}_2 + \dot{\theta}_3\right)^2 (a_3 + g_3) \\
\\
\ddot{\theta}_1 a_1\,\mathrm{c}\theta_{23} + \left(\ddot{\theta}_1 + \ddot{\theta}_2\right) a_2\,\mathrm{c}\theta_3 + \left(\ddot{\theta}_1 + \ddot{\theta}_2 + \ddot{\theta}_3\right)(a_3 + g_3) \\
+\dot{\theta}_1^2 a_1\,\mathrm{s}\theta_{23} - \left(\dot{\theta}_1 + \dot{\theta}_2\right)^2 a_1\,\mathrm{s}\theta_3 \\
\\
\varepsilon\left[\begin{array}{c}
\ddot{\theta}_1 a_1 g_3\,\mathrm{c}\theta_{23} + \left(\ddot{\theta}_1 + \ddot{\theta}_2\right) a_2 g_3\,\mathrm{c}\theta_3 + \left(\ddot{\theta}_1 + \ddot{\theta}_2 + \ddot{\theta}_3\right)\left(a_3 g_3 + k_{3z}^2\right) \\
-\dot{\theta}_2\dot{\theta}_3 a_2 g_3\,\mathrm{s}\theta_3 + \dot{\theta}_1\dot{\theta}_3 a_1 g_3\,\mathrm{s}\theta_3 + \left(\dot{\theta}_1 + \dot{\theta}_2\right)^2 a_2 g_3\,\mathrm{s}\theta_3
\end{array}\right]
\end{array}\right\} \tag{702}
$$

Eqs.(700, 701, 702) are evaluated to obtain the inertia forces to be substituted into the equations of dynamic equilibrium which will now be developed by application of D'Alembert's principle.

The joint dual forces are

$$
{}^{1}\hat{\mathrm{F}}_1 = \left\{\begin{array}{c} F_{1i} \\ F_{1j} \\ \varepsilon T_{1k} \end{array}\right\} \tag{703}
$$

$$
{}^{2}\hat{\mathrm{F}}_2 = \left\{\begin{array}{c} F_{2i} \\ F_{2j} \\ 0 \end{array}\right\} \tag{704}
$$

$$
{}^{3}\hat{\mathrm{F}}_3 = \left\{\begin{array}{c} F_{3i} \\ F_{3j} \\ 0 \end{array}\right\} \tag{705}
$$

$$
{}^{4}\hat{\mathrm{F}}_4 = \left\{\begin{array}{c} F_{4i} \\ F_{4j} \\ \varepsilon T_{4k} \end{array}\right\} \tag{706}
$$

where it is recognized that the intermediate revolute joints cannot support a torque about their axes.

The equation of dynamic equilibrium for the input crank is

$$
0 = {}^{2}_{1}\hat{\mathrm{M}}\,{}^{1}\hat{\mathrm{F}}_1 - {}^{2}\hat{\mathrm{F}}_2 - {}^{2}\hat{\mathrm{f}}_1 = \begin{bmatrix} \mathrm{c}\theta_1 & \mathrm{s}\theta_1 & 0 \\ -\mathrm{s}\theta_1 & \mathrm{c}\theta_1 & \varepsilon a_1 \\ \varepsilon a_1\,\mathrm{s}\theta_1 & -\varepsilon a_1\,\mathrm{c}\theta_1 & 1 \end{bmatrix}\left\{\begin{array}{c} F_{1i} \\ F_{1j} \\ \varepsilon T_{1k} \end{array}\right\} - \left\{\begin{array}{c} F_{2i} \\ F_{2j} \\ 0 \end{array}\right\} - \left\{\begin{array}{c} f_{1i} \\ f_{1j} \\ \varepsilon t_{1k} \end{array}\right\} \tag{707}
$$

where it has been observed from eq.(700) that inertia dual force components t_{1i}, t_{1j} and f_{1k} are zero.

$$0=\begin{Bmatrix} F_{1i}\,c\theta_1+F_{1j}\,s\theta_1-F_{2i}-f_{1i} \\ -F_{1i}\,s\theta_1+F_{1j}\,c\theta_1-F_{2j}-f_{1j} \\ \varepsilon\left[a_1\left(F_{1i}\,s\theta_1-F_{1j}\,c\theta_1\right)+T_{1k}-t_{1k}\right] \end{Bmatrix} \tag{708}$$

The equation of dynamic equilibrium for the coupler is

$$0={}_2^3\hat{M}\,{}^2\hat{F}_2-{}^3\hat{F}_3-{}^3\hat{f}_2 \tag{709}$$

$$=\begin{bmatrix} c\theta_2 & s\theta_2 & 0 \\ -s\theta_2 & c\theta_2 & \varepsilon a_2 \\ \varepsilon a_2\,s\theta_2 & -\varepsilon a_2\,c\theta_2 & 1 \end{bmatrix}\begin{Bmatrix} F_{2i} \\ F_{2j} \\ 0 \end{Bmatrix}-\begin{Bmatrix} F_{3i} \\ F_{3j} \\ 0 \end{Bmatrix}-\begin{Bmatrix} f_{2i} \\ f_{2j} \\ \varepsilon t_{2k} \end{Bmatrix} \tag{710}$$

where it has been observed from eq.(701) that inertia dual force components t_{2i}, t_{2j} and f_{2k} are zero.

$$0=\begin{Bmatrix} F_{2i}\,c\theta_2+F_{2j}\,s\theta_2-F_{3i}-f_{2i} \\ -F_{2i}\,s\theta_2+F_{2j}\,c\theta_2-F_{3j}-f_{2j} \\ \varepsilon\left[a_2\left(F_{2i}\,s\theta_2-F_{2j}\,c\theta_2\right)-t_{2k}\right] \end{Bmatrix} \tag{711}$$

and the equation of dynamic equilibrium for the output crank is

$$0={}_3^4\hat{M}\,{}^3\hat{F}_3-{}^4\hat{F}_4-{}^4\hat{f}_3 \tag{712}$$

$$=\begin{bmatrix} c\theta_3 & s\theta_3 & 0 \\ -s\theta_3 & c\theta_3 & \varepsilon a_3 \\ \varepsilon a_3\,s\theta_3 & -\varepsilon a_3\,c\theta_3 & 1 \end{bmatrix}\begin{Bmatrix} F_{3i} \\ F_{3j} \\ 0 \end{Bmatrix}-\begin{Bmatrix} F_{4i} \\ F_{4j} \\ \varepsilon T_{4k} \end{Bmatrix}-\begin{Bmatrix} f_{3i} \\ f_{3j} \\ \varepsilon t_{3k} \end{Bmatrix} \tag{713}$$

where it has been observed from eq.(702) that inertia dual force components t_{3i}, t_{3j} and f_{3k} are zero.

$$0=\begin{Bmatrix} F_{3i}\,c\theta_3+F_{3j}\,s\theta_3-F_{4i}-f_{3i} \\ -F_{3i}\,s\theta_3+F_{3j}\,c\theta_3-F_{4j}-f_{3j} \\ \varepsilon\left[a_3\left(F_{3i}\,s\theta_3-F_{3j}\,c\theta_3\right)-T_{4k}-t_{3k}\right] \end{Bmatrix} \tag{714}$$

Values of the displacements, together with the inertia forces calculated by eqs.(700, 701, 702), calculated on the basis of the link dimensions and masses and the absolute velocities, are substituted into the three dual vector equations, eqs.(708, 711, 714) to establish a set of nine non-trivial scalar equations which can be solved for the eight reaction forces and the output torque. This set is written in matrix form as

$$\begin{bmatrix} c\theta_1 & s\theta_1 & -1 & 0 & 0 & 0 & 0 & 0 & 0 \\ -s\theta_1 & c\theta_1 & 0 & -1 & 0 & 0 & 0 & 0 & 0 \\ a_1 s\theta_1 & -a_1 c\theta_1 & 0 & 0 & 0 & 0 & 0 & 0 & 0 \\ 0 & 0 & c\theta_2 & s\theta_2 & -1 & 0 & 0 & 0 & 0 \\ 0 & 0 & -s\theta_2 & c\theta_2 & 0 & -1 & 0 & 0 & 0 \\ 0 & 0 & a_2 s\theta_2 & -a_2 c\theta_2 & 0 & 0 & 0 & 0 & 0 \\ 0 & 0 & 0 & 0 & c\theta_3 & s\theta_3 & -1 & 0 & 0 \\ 0 & 0 & 0 & 0 & -s\theta_3 & c\theta_3 & 0 & -1 & 0 \\ 0 & 0 & 0 & 0 & a_3 s\theta_3 & -a_3 c\theta_3 & 0 & 0 & -1 \end{bmatrix} \begin{Bmatrix} F_{1i} \\ F_{1j} \\ F_{2i} \\ F_{2j} \\ F_{3i} \\ F_{3j} \\ F_{4i} \\ F_{4j} \\ T_{4k} \end{Bmatrix} = \begin{Bmatrix} f_{1i} \\ f_{1j} \\ t_{1k} - T_{1k} \\ f_{2i} \\ f_{2j} \\ t_{2k} \\ f_{3i} \\ f_{3j} \\ t_{3k} \end{Bmatrix} \tag{715}$$

and has solutions for the joint reaction force components and the output-shaft torque

$$F_{1i} = \left[\frac{\left(T_{1k} - t_{1k}\right)}{a_1} c\theta_{12} + \left(\frac{t_{2k}}{a_2} - f_{1j} c\theta_2\right) c\theta_1\right] \frac{1}{s\theta_2} + f_{1i} c\theta_1 \tag{716}$$

$$F_{1j} = \left[\frac{\left(T_{1k} - t_{1k}\right)}{a_1} s\theta_{12} + \left(\frac{t_{2k}}{a_2} - f_{1j} c\theta_2\right) s\theta_1\right] \frac{1}{s\theta_2} + f_{1i} s\theta_1 \tag{717}$$

$$F_{2i} = \left[\frac{\left(T_{1k} - t_{1k}\right)}{a_1} c\theta_2 + \frac{t_{2k}}{a_2} - f_{1j} c\theta_2\right] \frac{1}{s\theta_2} \tag{718}$$

$$F_{2j} = \frac{T_{1k} - t_{1k}}{a_1} - f_{1j} \tag{719}$$

$$F_{3i} = \left[\frac{\left(T_{1k} - t_{1k}\right)}{a_1} + \frac{t_{2k}}{a_2} c\theta_2 - f_{1j}\right] \frac{1}{s\theta_2} - f_{2j} \tag{720}$$

$$F_{3j} = -\frac{t_{2k}}{a_2} - f_{2j} \tag{721}$$

$$F_{4i} = \left\{\left[\frac{\left(T_{1k} - t_{1k}\right)}{a_1} - f_{1j}\right] c\theta_3 + \frac{t_{2k}}{a_2} c\theta_{23}\right\} \frac{1}{s\theta_2} - f_{2i} c\theta_3 - f_{2j} s\theta_3 - f_{3i} \tag{722}$$

$$F_{4i} = -\left\{\left[\frac{\left(T_{1k} - t_{1k}\right)}{a_1} - f_{1j}\right] s\theta_3 + \frac{t_{2k}}{a_2} s\theta_{23}\right\} \frac{1}{s\theta_2} - f_{2i} s\theta_3 - f_{2j} c\theta_3 - f_{3j} \tag{723}$$

$$T_{4k} = a_3\left\{\left[\frac{(T_{1k} - t_{1k})}{a_1} - f_{1j}\right]\frac{s\theta_3}{s\theta_2} + \frac{t_{2k}}{a_2}\frac{s\theta_{23}}{s\theta_3} - f_{2i}\,s\theta_3 + f_{2j}\,c\theta_3\right\} - t_{3k} \qquad (724)$$

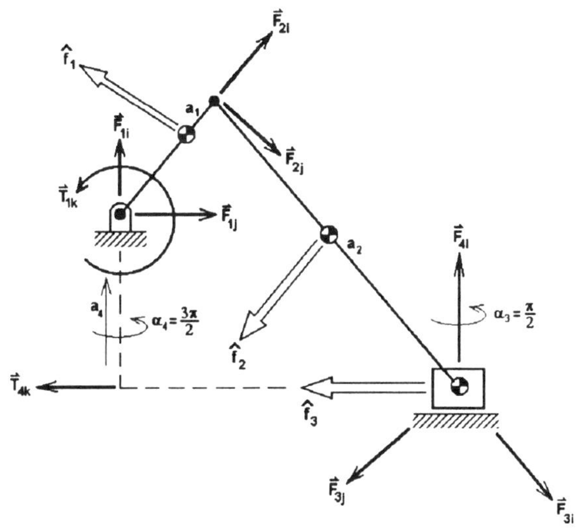

Figure 61 Dynamic loading of the planar slider-crank mechanism.

Planar Slider-Crank Mechanism

The absolute velocities of each moving link of a planar slider-crank mechanism as illustrated in Figure 61 in terms of the frames fixed at the distal ends of each link are

$$^2\hat{V}^2_{14} = \begin{Bmatrix} 0 \\ \varepsilon\dot{\theta}_1 a_1 \\ \dot{\theta}_1 \end{Bmatrix} \qquad (725)$$

$$^3\hat{V}^3_{24} = \begin{Bmatrix} \varepsilon\dot{\theta}_1 a_1\, s\theta_2 \\ \varepsilon\left[\dot{\theta}_1(a_1\, c\theta_2 + a_2) + \dot{\theta}_2 a_2\right] \\ \dot{\theta}_1 + \dot{\theta}_2 \end{Bmatrix} \qquad (726)$$

$$^4\hat{V}^4_{34} = \begin{Bmatrix} 0 \\ 0 \\ -\varepsilon\dot{s}_4 \end{Bmatrix} \qquad (727)$$

It will be assumed that the input crank and connecting rod are symmetrical about their lengths and that the center of mass of the slider is located at the wrist pin so that the mass factors can be simplified. Thus,

$$^{2}\vec{G}_1 = \begin{Bmatrix} g_1 \\ 0 \\ 0 \end{Bmatrix}, \qquad ^{3}\vec{G}_2 = \begin{Bmatrix} g_2 \\ 0 \\ 0 \end{Bmatrix}, \qquad ^{4}\vec{G}_3 = \begin{Bmatrix} g_3 \\ 0 \\ 0 \end{Bmatrix} \tag{728}$$

and so the first moments of inertia are

$$\left[{}^{2}S_1^{2}\right] = m_1 \begin{bmatrix} 0 & 0 & 0 \\ 0 & 0 & -g_1 \\ 0 & g_1 & 0 \end{bmatrix} \tag{729}$$

$$\left[{}^{3}S_2^{3}\right] = m_2 \begin{bmatrix} 0 & 0 & 0 \\ 0 & 0 & -g_2 \\ 0 & g_2 & 0 \end{bmatrix} \tag{730}$$

$$\left[{}^{4}S_3^{4}\right] = m_3 \begin{bmatrix} 0 & 0 & 0 \\ 0 & 0 & 0 \\ 0 & 0 & 0 \end{bmatrix} \tag{731}$$

It will also be assumed that the principal axes of each link are oriented the same as the frame fixed on the distal end of each link. If we let symbols k_{Ax}, k_{Ay}, k_{Az} represent the radii of gyration of each link A, then the second moments of inertia of each link in terms of coordinate frames fixed on the ends of the links are

$$\left[{}^{2}J_1^{2}\right] = m_1 \begin{bmatrix} k_{1x}^2 & 0 & 0 \\ 0 & k_{1y}^2 & 0 \\ 0 & 0 & k_{1z}^2 \end{bmatrix} \tag{732}$$

$$\left[{}^{3}J_2^{3}\right] = m_2 \begin{bmatrix} k_{2x}^2 & 0 & 0 \\ 0 & k_{2y}^2 & 0 \\ 0 & 0 & k_{2z}^2 \end{bmatrix} \tag{733}$$

$$\left[{}^{4}J_3^{4}\right] = m_3 \begin{bmatrix} k_{3x}^2 & 0 & 0 \\ 0 & k_{3y}^2 & 0 \\ 0 & 0 & k_{3z}^2 \end{bmatrix} \tag{734}$$

Now, the dual momentum of link A in terms of frame {B} fixed on its distal end is

$$ {}^{B}\hat{\mathrm{H}}_{A} = m_{A}\,{}^{B}\vec{\mathrm{V}}_{Ad}^{B} - \left[{}^{B}S_{A}^{B}\right]{}^{B}\vec{\mathrm{V}}_{Ap}^{B} + \varepsilon\left(\left[{}^{B}S_{A}^{B}\right]{}^{B}\vec{\mathrm{V}}_{Ad}^{B} + \left[{}^{B}J_{A}^{B}\right]{}^{B}\vec{\mathrm{V}}_{Ap}^{B}\right) \quad (735) $$

So, for the input crank, substituting eqs.(725, 729, 732) into eq.(735),

$$ {}^{2}\hat{\mathrm{H}}_{1} = m_{1}\begin{Bmatrix} 0 \\ \dot{\theta}_{1}(a_{1}+g_{1}) \\ \varepsilon\dot{\theta}_{1}\left(a_{1}g_{1}+k_{1z}^{2}\right) \end{Bmatrix} \quad (736) $$

and for the connecting rod, substituting eqs.(726, 730, 733) into eq.(735),

$$ {}^{3}\hat{\mathrm{H}}_{2} = m_{2}\begin{Bmatrix} \dot{\theta}_{1}a_{1}\,\mathrm{s}\theta_{2} \\ \dot{\theta}_{1}a_{1}\,\mathrm{c}\theta_{2} + \left(\dot{\theta}_{1}+\dot{\theta}_{2}\right)(a_{2}+g_{2}) \\ \varepsilon\left[\dot{\theta}_{1}a_{1}g_{2}\,\mathrm{c}\theta_{2} + \left(\dot{\theta}_{1}+\dot{\theta}_{2}\right)\left(a_{2}g_{2}+k_{1z}^{2}\right)\right] \end{Bmatrix} \quad (737) $$

and for the slider, subsituting eqs.(727, 731, 734) into eq.(735),

$$ {}^{4}\hat{\mathrm{H}}_{3} = m_{3}\begin{Bmatrix} 0 \\ 0 \\ -\dot{s}_{4} \end{Bmatrix} \quad (738) $$

The time derivatives of the dual momenta of the links are obtained by differentiation of eqs.(736, 737, 738) as

$$ {}^{2}\dot{\hat{\mathrm{H}}}_{1} = m_{1}\begin{Bmatrix} 0 \\ \ddot{\theta}_{1}(a_{1}+g_{1}) \\ \varepsilon\ddot{\theta}_{1}\left(a_{1}g_{1}+k_{1z}^{2}\right) \end{Bmatrix} \quad (739) $$

$$ {}^{3}\dot{\hat{\mathrm{H}}}_{2} = m_{2}\begin{Bmatrix} \ddot{\theta}_{1}a_{1}\,\mathrm{s}\theta_{2} + \dot{\theta}_{1}\dot{\theta}_{2}a_{1}\,\mathrm{c}\theta_{2} \\ \ddot{\theta}_{1}a_{1}\,\mathrm{c}\theta_{2} - \dot{\theta}_{1}\dot{\theta}_{2}a_{1}\,\mathrm{s}\theta_{2} + \left(\ddot{\theta}_{1}+\ddot{\theta}_{2}\right)(a_{2}+g_{2}) \\ \varepsilon\left[\ddot{\theta}_{1}a_{1}g_{2}\,\mathrm{c}\theta_{2} - \dot{\theta}_{1}\dot{\theta}_{2}a_{1}g_{2}\,\mathrm{s}\theta_{2} + \left(\ddot{\theta}_{1}+\ddot{\theta}_{2}\right)\left(a_{2}g_{2}+k_{1z}^{2}\right)\right] \end{Bmatrix} \quad (740) $$

$$ {}^{4}\dot{\hat{\mathrm{H}}}_{3} = m_{3}\begin{Bmatrix} 0 \\ 0 \\ -\ddot{s}_{4} \end{Bmatrix} \quad (741) $$

Next we evaluate the inertia dual forces. For the input crank, substituting eqs.(725, 736, 739) into the dual Euler equation

$$
{}^{2}\hat{\mathrm{f}}_1 = \begin{Bmatrix} {}^{2}\hat{f}_{1i} \\ {}^{2}\hat{f}_{1j} \\ {}^{2}\hat{f}_{1k} \end{Bmatrix} = \begin{Bmatrix} {}^{2}\dot{\hat{H}}_{1i} - {}^{2}\hat{V}_{1k}^{22}\,{}^{2}\hat{H}_{1j} + {}^{2}\hat{V}_{1j}^{22}\,{}^{2}\hat{H}_{1k} \\ {}^{2}\dot{\hat{H}}_{1j} - {}^{2}\hat{V}_{1i}^{22}\,{}^{2}\hat{H}_{1k} + {}^{2}\hat{V}_{1k}^{22}\,{}^{2}\hat{H}_{1i} \\ {}^{2}\dot{\hat{H}}_{1k} - {}^{2}\hat{V}_{1j}^{22}\,{}^{2}\hat{H}_{1i} + {}^{2}\hat{V}_{1i}^{22}\,{}^{2}\hat{H}_{1j} \end{Bmatrix} \tag{742}
$$

yields

$$
{}^{2}\hat{\mathrm{f}}_1 = m_1 \begin{Bmatrix} \dot{\theta}_1^2 (a_1 + g_1) \\ \ddot{\theta}_1 (a_1 + g_1) \\ \varepsilon \ddot{\theta}_1 (a_1 g_1 + k_{1z}^2) \end{Bmatrix} \tag{743}
$$

and substituting eqs.(726, 737, 740) into the dual Euler equation for the connecting rod

$$
{}^{3}\hat{\mathrm{f}}_2 = \begin{Bmatrix} {}^{3}\hat{f}_{2i} \\ {}^{3}\hat{f}_{2j} \\ {}^{3}\hat{f}_{2k} \end{Bmatrix} = \begin{Bmatrix} {}^{3}\dot{\hat{H}}_{2i} - {}^{3}\hat{V}_{2k}^{33}\,{}^{3}\hat{H}_{2j} + {}^{3}\hat{V}_{2j}^{33}\,{}^{3}\hat{H}_{2k} \\ {}^{3}\dot{\hat{H}}_{2j} - {}^{3}\hat{V}_{2i}^{33}\,{}^{3}\hat{H}_{2k} + {}^{3}\hat{V}_{2k}^{33}\,{}^{3}\hat{H}_{2i} \\ {}^{3}\dot{\hat{H}}_{2k} - {}^{3}\hat{V}_{2j}^{33}\,{}^{3}\hat{H}_{2i} + {}^{3}\hat{V}_{2i}^{33}\,{}^{3}\hat{H}_{2j} \end{Bmatrix} \tag{744}
$$

yields

$$
{}^{3}\hat{\mathrm{f}}_2 = m_2 \begin{Bmatrix} \ddot{\theta}_1 a_1\, \mathrm{s}\theta_2 - \dot{\theta}_1^2 a_1\, \mathrm{c}\theta_2 + (\dot{\theta}_1 + \dot{\theta}_2)^2 (a_2 + g_2) \\ \dot{\theta}_1 a_1\, \mathrm{c}\theta_2 + (\ddot{\theta}_1 + \ddot{\theta}_2)(a_2 + g_2) + \dot{\theta}_1^2 a_1\, \mathrm{s}\theta_2 \\ \varepsilon\left[\ddot{\theta}_1 a_1 g_2\, \mathrm{c}\theta_2 + (\ddot{\theta}_1 + \ddot{\theta}_2)(a_2 g_2 + k_{1z}^2) - \dot{\theta}_1^2 a_1 g_2\, \mathrm{s}\theta_2\right] \end{Bmatrix} \tag{745}
$$

and substituting eqs.(727, 738, 741) into the dual Euler equation for the slider

$$
{}^{4}\hat{\mathrm{f}}_3 = \begin{Bmatrix} {}^{4}\hat{f}_{3i} \\ {}^{4}\hat{f}_{3j} \\ {}^{4}\hat{f}_{3k} \end{Bmatrix} = \begin{Bmatrix} {}^{4}\dot{\hat{H}}_{3i} - {}^{4}\hat{V}_{3k}^{44}\,{}^{4}\hat{H}_{3j} + {}^{4}\hat{V}_{3j}^{44}\,{}^{4}\hat{H}_{3k} \\ {}^{4}\dot{\hat{H}}_{3j} - {}^{4}\hat{V}_{3i}^{44}\,{}^{4}\hat{H}_{3k} + {}^{4}\hat{V}_{3k}^{44}\,{}^{4}\hat{H}_{3i} \\ {}^{4}\dot{\hat{H}}_{3k} - {}^{4}\hat{V}_{3j}^{44}\,{}^{4}\hat{H}_{3i} + {}^{3}\hat{V}_{3i}^{44}\,{}^{4}\hat{H}_{3j} \end{Bmatrix} \tag{746}
$$

results in

$$
{}^{4}\hat{\mathrm{f}}_3 = m_3 \begin{Bmatrix} 0 \\ 0 \\ -\ddot{s}_4 \end{Bmatrix} \tag{747}
$$

Values of the dual inertia forces calculated by use of eqs.(743, 745, 747) can be substituted into the equations of dynamic equilibrium which will now be developed.

Now, the dual force at joint 1 is

$$ {}^{1}\hat{\mathrm{F}}_{1} = \begin{Bmatrix} F_{1i} \\ F_{1j} \\ \varepsilon T_{1k} \end{Bmatrix} \tag{748} $$

where torque component T_{1k} is the applied input.

At joints 2 and 3,

$$ {}^{2}\hat{\mathrm{F}}_{2} = \begin{Bmatrix} F_{2i} \\ F_{2j} \\ 0 \end{Bmatrix} \tag{749} $$

$$ {}^{3}\hat{\mathrm{F}}_{3} = \begin{Bmatrix} F_{3i} \\ F_{3j} \\ 0 \end{Bmatrix} \tag{750} $$

where it has been noted that an intermediate revolute joint cannot support a torque about its axis.

At joint 4,

$$ {}^{4}\hat{\mathrm{F}}_{4} = \begin{Bmatrix} F_{4i} \\ 0 \\ F_{4k} \end{Bmatrix} \tag{751} $$

where component F_{4k} represents the output force.

The equation of dynamic equilibrium of each moving link A can be obtained by application of the dual Euler equation.

$$ {}^{B}_{A}\hat{\mathrm{M}}\, {}^{A}\hat{\mathrm{F}}_{A} - {}^{B}\hat{\mathrm{F}}_{B} = {}^{B}\hat{\mathrm{f}}_{A} \tag{752} $$

For the input crank, writing eq.(752)as

$$ {}^{2}_{1}\hat{\mathrm{M}}\, {}^{1}\hat{\mathrm{F}}_{1} - {}^{2}\hat{\mathrm{F}}_{2} = {}^{2}\hat{\mathrm{f}}_{1} \tag{753} $$

results in

$$\begin{bmatrix} c\theta_1 & s\theta_1 & 0 \\ -s\theta_1 & c\theta_1 & \varepsilon a_1 \\ \varepsilon a_1 s\theta_1 & -\varepsilon a_1 c\theta_1 & 0 \end{bmatrix} \begin{Bmatrix} F_{1i} \\ F_{1j} \\ \varepsilon T_{1k} \end{Bmatrix} - \begin{Bmatrix} F_{2i} \\ F_{2j} \\ 0 \end{Bmatrix} = \begin{Bmatrix} f_{1i} \\ f_{1j} \\ \varepsilon t_{1k} \end{Bmatrix} \tag{754}$$

$$\begin{Bmatrix} F_{1i} c\theta_1 + F_{1j} s\theta_1 - F_{2i} \\ -F_{1i} s\theta_1 + F_{1j} c\theta_1 - F_{2j} \\ \varepsilon\left[F_{1i} a_1 s\theta_1 - F_{1j} a_1 c\theta_1 + T_{1k}\right] \end{Bmatrix} = \begin{Bmatrix} f_{1i} \\ f_{1j} \\ \varepsilon t_{1k} \end{Bmatrix} \tag{755}$$

For the connecting rod, eq.(752) is written as

$$^{3}_{2}\hat{\mathrm{M}}\,{}^{2}\hat{\mathrm{F}}_2 - {}^{3}\hat{\mathrm{F}}_3 = {}^{3}\hat{\mathrm{f}}_2 \tag{756}$$

and so

$$\begin{bmatrix} c\theta_2 & s\theta_2 & 0 \\ -s\theta_2 & c\theta_2 & \varepsilon a_2 \\ \varepsilon a_2 s\theta_2 & -\varepsilon a_2 c\theta_2 & 0 \end{bmatrix} \begin{Bmatrix} F_{2i} \\ F_{2j} \\ \varepsilon T_{2k} \end{Bmatrix} - \begin{Bmatrix} F_{2i} \\ F_{2j} \\ 0 \end{Bmatrix} = \begin{Bmatrix} f_{2i} \\ f_{2j} \\ \varepsilon t_{2k} \end{Bmatrix} \tag{757}$$

$$\begin{Bmatrix} F_{2i} c\theta_2 + F_{2j} s\theta_2 - F_{3i} \\ -F_{2i} s\theta_2 + F_{2j} c\theta_2 - F_{3j} \\ \varepsilon\left[F_{2i} a_2 s\theta_2 - F_{2j} a_2 c\theta_2\right] \end{Bmatrix} = \begin{Bmatrix} f_{2i} \\ f_{2j} \\ \varepsilon t_{2k} \end{Bmatrix} \tag{758}$$

For the slider, eq.(752) becomes

$$^{4}_{3}\hat{\mathrm{M}}\,{}^{3}\hat{\mathrm{F}}_3 - {}^{4}\hat{\mathrm{F}}_4 = {}^{4}\hat{\mathrm{f}}_3 \tag{759}$$

$$\begin{bmatrix} c\theta_3 & s\theta_3 & 0 \\ 0 & 0 & 1 \\ s\theta_3 & -c\theta_3 & 0 \end{bmatrix} \begin{Bmatrix} F_{3i} \\ F_{3j} \\ 0 \end{Bmatrix} - \begin{Bmatrix} F_{4i} \\ 0 \\ F_{4k} \end{Bmatrix} = \begin{Bmatrix} 0 \\ 0 \\ f_{3k} \end{Bmatrix} \tag{760}$$

$$\begin{Bmatrix} F_{3i} c\theta_3 + F_{3j} s\theta_3 - F_{4i} \\ 0 \\ F_{3i} s\theta_3 - F_{3j} c\theta_3 - F_{4k} \end{Bmatrix} = \begin{Bmatrix} 0 \\ 0 \\ f_{3k} \end{Bmatrix} \tag{761}$$

Eqs.(755, 757, 761) constitute the equations of dynamic equilibrium for the planar slider-crank mechanism and can be combined in matrix form as a system of eight equations linear in eight unknown forces,

$$\begin{bmatrix} c\theta_1 & s\theta_1 & -1 & 0 & 0 & 0 & 0 & 0 \\ -s\theta_1 & c\theta_1 & 0 & -1 & 0 & 0 & 0 & 0 \\ a_1 s\theta_1 & -a_1 c\theta_1 & 0 & 0 & 0 & 0 & 0 & 0 \\ 0 & 0 & c\theta_2 & s\theta_2 & -1 & 0 & 0 & 0 \\ 0 & 0 & -s\theta_2 & c\theta_2 & 0 & -1 & 0 & 0 \\ 0 & 0 & a_2 s\theta_2 & a_2 c\theta_2 & 0 & 0 & 0 & 0 \\ 0 & 0 & 0 & 0 & c\theta_3 & s\theta_3 & -1 & 0 \\ 0 & 0 & 0 & 0 & s\theta_3 & -c\theta_3 & 0 & -1 \end{bmatrix} \begin{Bmatrix} F_{1i} \\ F_{1j} \\ F_{2i} \\ F_{2j} \\ F_{3i} \\ F_{3j} \\ F_{4i} \\ F_{4k} \end{Bmatrix} = \begin{Bmatrix} f_{1i} \\ f_{1j} \\ t_{1k} - T_{1k} \\ f_{2i} \\ f_{2j} \\ t_{2k} \\ 0 \\ f_{3k} \end{Bmatrix} \quad (762)$$

The solutions for the unknown reaction force components and the unknown output force are

$$F_{1i} = \frac{(T_{1k} - t_{1k})\, c\theta_{12}}{a_1\, s\theta_2} - \left[\frac{(a_2 f_{1j}\, c\theta_2 + t_{2k})}{a_2\, s\theta_2} + f_{1i} \right] c\theta_1 \quad (763)$$

$$F_{1j} = \frac{(T_{1k} - t_{1k})\, s\theta_{12}}{a_1\, s\theta_2} \quad (764)$$

$$F_{2i} = \frac{T_{1k} - t_{1k}}{a_2 \tan\theta_2} + \frac{t_{2k}}{a_2\, s\theta_2} - \frac{f_{1j}}{\tan\theta_2} \quad (765)$$

$$F_{2j} = \frac{T_{1k} - t_{1k}}{a_1} - f_{1j} \quad (766)$$

$$F_{3i} = \frac{T_{1k} - t_{1k}}{a_1\, s\theta_2} + \frac{t_{2k}\, c\theta_2}{a_2\, s\theta_2} - \frac{f_{1j}}{s\theta_2} + f_{2j} \quad (767)$$

$$F_{3j} = -\frac{t_{2k}}{a_2} - f_{2j} \quad (768)$$

$$F_{4i} = \frac{(T_{1k} - t_{1k} - a_1 f_{1j})\, c\theta_3}{a_1\, s\theta_2} + \frac{t_{2k}\, c\theta_{23}}{a_2\, s\theta_2} - f_{2i}\, c\theta_3 - f_{2j}\, s\theta_3 \quad (769)$$

$$F_{4k} = \frac{(T_{1k} - t_{1k} - a_1 f_{1j})\, s\theta_3}{a_1\, s\theta_2} + \frac{t_{2k} s\theta_{23}}{a_2\, s\theta_2} - f_{2i}\, s\theta_3 - f_{2j}\, c\theta_3 - f_{3k} \quad (770)$$

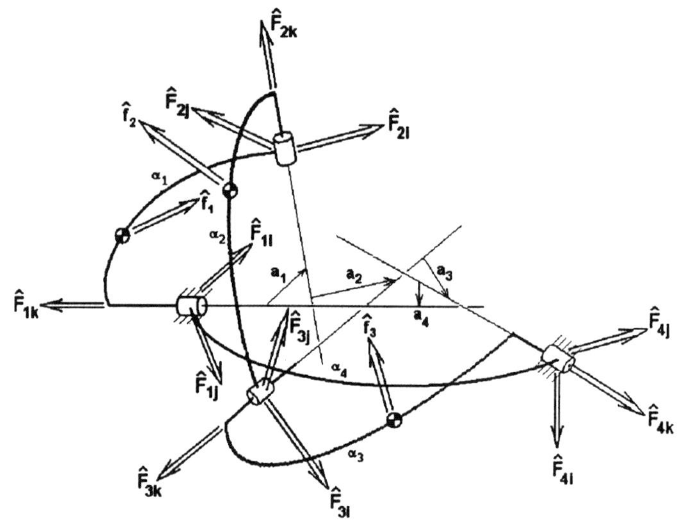

Figure 62 Dynamic loading of the spatial four-bar mechanism.

Spatial Four-Bar Mechanism

The absolute velocities of the links of a spatial four-bar mechanism, as shown in Figure 62, are

$$ {}^{2}V_{14}^{2} = \vec{\omega}_1 + \varepsilon \vec{v}_1 = \left\{ \begin{matrix} 0 \\ \dot{\theta}_n s\alpha_1 \\ \dot{\theta}_n c\alpha_1 \end{matrix} \right\} + \varepsilon \left\{ \begin{matrix} 0 \\ \dot{\theta}_1 a_1 c\alpha_1 + \dot{s}_1 s\alpha_1 \\ -\dot{\theta}_1 a_1 s\alpha_1 + \dot{s}_1 c\alpha_1 \end{matrix} \right\} \tag{771} $$

$$ {}^{3}V_{24}^{3} = \vec{\omega}_2 + \varepsilon \vec{v}_2 = \left\{ \begin{matrix} -\dot{\theta}_4 s\alpha_3 s\theta_3 \\ \dot{\theta}_4 s\alpha_3 c\theta_3 \\ -\dot{\theta}_3 + \dot{\theta}_4 c\alpha_3 \end{matrix} \right\} + \varepsilon \left\{ \begin{matrix} -\dot{\theta}_4 (a_3 c\alpha_3 s\theta_3 + s_3 s\alpha_3 c\theta_3) - \dot{s}_4 s\alpha_3 s\theta_3 \\ \dot{\theta}_4 (a_3 c\alpha_3 c\theta_3 - s_3 s\alpha_3 s\theta_3) - \dot{s}_4 s\alpha_3 c\theta_3 \\ -\dot{\theta}_4 a_3 s\alpha_3 - \dot{s}_3 + \dot{s}_4 c\alpha_3 \end{matrix} \right\} \tag{772} $$

$$ {}^{4}V_{34}^{4} = \vec{\omega}_4 + \varepsilon \vec{v}_4 = \left\{ \begin{matrix} 0 \\ 0 \\ -\dot{\theta}_4 \end{matrix} \right\} + \varepsilon \left\{ \begin{matrix} 0 \\ 0 \\ -\dot{s}_4 \end{matrix} \right\} \tag{773} $$

In a spatial mechanism the links are not typically rod-like and so it is not practical to make the simplifying assumptions about mass distribution which are reasonable for planar mechanisms. Consequently, all of the terms in the mass matrices will be carried. The first moments of inertia are

$$\left[{}^{2}S_{1}^{2}\right]=\begin{bmatrix} 0 & -S_{1z} & S_{1y} \\ S_{1z} & 0 & -S_{1x} \\ -S_{1y} & S_{1x} & 0 \end{bmatrix} \tag{774}$$

$$\left[{}^{3}S_{2}^{3}\right]=\begin{bmatrix} 0 & -S_{2z} & S_{2y} \\ S_{2z} & 0 & -S_{2x} \\ -S_{2y} & S_{2x} & 0 \end{bmatrix} \tag{775}$$

$$\left[{}^{4}S_{3}^{4}\right]=\begin{bmatrix} 0 & -S_{3z} & S_{3y} \\ S_{3z} & 0 & -S_{3x} \\ -S_{3y} & S_{3x} & 0 \end{bmatrix} \tag{776}$$

and the second moments of inertia are

$$\left[{}^{2}J_{1}^{2}\right]=\begin{bmatrix} J_{1xx} & -J_{1xy} & -J_{1xz} \\ -J_{1xy} & J_{1yy} & -J_{1yz} \\ -J_{1xz} & -J_{1yz} & J_{1zz} \end{bmatrix} \tag{777}$$

$$\left[{}^{3}J_{2}^{3}\right]=\begin{bmatrix} J_{2xx} & -J_{2xy} & -J_{2xz} \\ -J_{2xy} & J_{2yy} & -J_{2yz} \\ -J_{2xz} & -J_{2yx} & J_{2zz} \end{bmatrix} \tag{778}$$

$$\left[{}^{4}J_{3}^{4}\right]=\begin{bmatrix} J_{3xx} & -J_{3xy} & -J_{3xz} \\ -J_{3xy} & J_{3yy} & -J_{3yz} \\ -J_{3xz} & -J_{3yz} & J_{3zz} \end{bmatrix} \tag{779}$$

The dual momentum of the input crank is

$${}^{2}\hat{\mathrm{H}}_{1}=m_{1}\vec{\mathrm{v}}_{1}-\left[{}^{2}S_{1}^{2}\right]\vec{\omega}_{1}+\varepsilon\left(\left[{}^{2}S_{1}^{2}\right]\vec{\mathrm{v}}_{1}+\left[{}^{2}J_{1}^{2}\right]\vec{\omega}_{1}\right) \tag{780}$$

$$= \tag{781}$$

$$m_{1}\begin{Bmatrix} 0 \\ \dot{\theta}_{1}a_{1}\,\mathrm{c}\alpha_{1}+\dot{s}_{1}\,\mathrm{s}\alpha_{1} \\ -\dot{\theta}_{1}a_{1}\,\mathrm{s}\alpha_{1}+\dot{s}_{1}\,\mathrm{c}\alpha_{1} \end{Bmatrix}-\begin{bmatrix} 0 & -S_{1z} & S_{1y} \\ S_{1z} & 0 & -S_{1x} \\ -S_{1y} & S_{1x} & 0 \end{bmatrix}\begin{Bmatrix} 0 \\ \dot{\theta}_{1}\,\mathrm{s}\alpha_{1} \\ \dot{\theta}_{1}\,\mathrm{c}\alpha_{1} \end{Bmatrix}$$

$$+$$

$$\varepsilon\left(\begin{bmatrix} 0 & -S_{1z} & S_{1y} \\ S_{1z} & 0 & -S_{1x} \\ -S_{1y} & S_{1x} & 0 \end{bmatrix}\begin{Bmatrix} 0 \\ \dot{\theta}_{1}a_{1}\,\mathrm{c}\alpha_{1}+\dot{s}_{1}\,\mathrm{s}\alpha_{1} \\ -\dot{\theta}_{1}a_{1}\,\mathrm{s}\alpha_{1}+\dot{s}_{1}\,\mathrm{c}\alpha_{1} \end{Bmatrix}+\begin{bmatrix} J_{1xx} & -J_{1xy} & -J_{1xz} \\ -J_{1xy} & J_{1yy} & -J_{1yz} \\ -J_{1xz} & -J_{1yz} & J_{1zz} \end{bmatrix}\begin{Bmatrix} 0 \\ \dot{\theta}_{1}\,\mathrm{s}\alpha_{1} \\ \dot{\theta}_{1}\,\mathrm{c}\alpha_{1} \end{Bmatrix}\right)$$

$$
\begin{aligned}
&{}^2\hat{\mathrm{H}}_1 \\
&= \\
&{}^2\vec{\mathrm{P}}_1 + \varepsilon\, {}^2\vec{\mathrm{H}}_1 \\
&= \\
&\begin{Bmatrix} \dot{\theta}_1\left(S_{1z}\,\mathrm{s}\alpha_1 - S_{1y}\,\mathrm{c}\alpha_1\right) \\ \dot{\theta}_1\left(m_1 a_1 + S_{1x}\right)\mathrm{c}\alpha_1 + m_1\dot{s}_1\,\mathrm{s}\alpha_1 \\ -\dot{\theta}_1\left(m_1 a_1 + S_{1x}\right)\mathrm{s}\alpha_1 + m_1\dot{s}_1\,\mathrm{c}\alpha_1 \end{Bmatrix} \\
&+ \\
&\varepsilon\begin{Bmatrix} -\dot{\theta}_1\left[a_1\left(S_{1z}\,\mathrm{c}\alpha_1 + S_{1y}\,\mathrm{s}\alpha_1\right) + J_{1xy} + J_{1xz}\right] + \dot{s}_1\left(S_{1y}\,\mathrm{c}\alpha_1 - S_{1z}\,\mathrm{s}\alpha_1\right) \\ -\dot{\theta}_1\left(S_{1x}a_1\,\mathrm{s}\alpha_1 - J_{1yy} + J_{1yz}\right) - \dot{s}_1 S_{1x}\,\mathrm{c}\alpha_1 \\ \dot{\theta}_1\left(S_{1x}a_1\,\mathrm{c}\alpha_1 - J_{1yz} + J_{1zz}\right) + \dot{s}_1 S_{1x}\,\mathrm{s}\alpha_1 \end{Bmatrix}
\end{aligned}
\tag{782}
$$

and the dual momentum of the coupler is

$$
{}^3\hat{\mathrm{H}}_2 = m_2\vec{\mathrm{v}}_2 - \left[{}^3S_2^3\right]\vec{\omega}_2 + \varepsilon\left(\left[{}^3S_2^3\right]\vec{\mathrm{v}}_2 + \left[{}^3J_2^3\right]\vec{\omega}_2\right) \tag{783}
$$

$$
\begin{aligned}
&= \\
&m_2\begin{Bmatrix} -\dot{\theta}_4\left(a_3\,\mathrm{c}\alpha_3\,\mathrm{s}\theta_3 + s_3\,\mathrm{s}\alpha_3\,\mathrm{c}\theta_3\right) - \dot{s}_4\,\mathrm{s}\alpha_3\,\mathrm{s}\theta_3 \\ \dot{\theta}_4\left(a_3\,\mathrm{c}\alpha_3\,\mathrm{c}\theta_3 - s_3\,\mathrm{s}\alpha_3\,\mathrm{s}\theta_3\right) - \dot{s}_4\,\mathrm{s}\alpha_3\,\mathrm{c}\theta_3 \\ \dot{s}_4\,\mathrm{c}\alpha_3 - \dot{s}_3 - \dot{\theta}_4 a_3\,\mathrm{s}\alpha_3 \end{Bmatrix} \\
&-\begin{bmatrix} 0 & -S_{2z} & S_{2y} \\ S_{2z} & 0 & -S_{2x} \\ -S_{2y} & S_{2x} & 0 \end{bmatrix}\begin{Bmatrix} -\dot{\theta}_4\,\mathrm{s}\alpha_3\,\mathrm{s}\theta_3 \\ \dot{\theta}_4\,\mathrm{s}\alpha_3\,\mathrm{c}\theta_3 \\ -\dot{\theta}_3 + \dot{\theta}_4\,\mathrm{c}\alpha_3 \end{Bmatrix} \\
&+\varepsilon\left(\begin{aligned} &\begin{bmatrix} 0 & -S_{2z} & S_{2y} \\ S_{2z} & 0 & -S_{2x} \\ -S_{2y} & S_{2x} & 0 \end{bmatrix}\begin{Bmatrix} -\dot{\theta}_4\left(a_3\,\mathrm{c}\alpha_3\,\mathrm{s}\theta_3 + s_3\,\mathrm{s}\alpha_3\,\mathrm{c}\theta_3\right) - \dot{s}_4\,\mathrm{s}\alpha_3\,\mathrm{s}\theta_3 \\ \dot{\theta}_4\left(a_3\,\mathrm{c}\alpha_3\,\mathrm{c}\theta_3 + s_3\,\mathrm{s}\alpha_3\,\mathrm{s}\theta_3\right) + \dot{s}_4\,\mathrm{s}\alpha_3\,\mathrm{c}\theta_3 \\ \dot{s}_4\,\mathrm{c}\alpha_3 - \dot{s}_3 - \dot{\theta}_4 a_3\,\mathrm{s}\alpha_3 \end{Bmatrix} \\ &\quad + \begin{bmatrix} J_{2xx} & -J_{2xy} & -J_{2xz} \\ -J_{2xy} & J_{2yy} & -J_{2yz} \\ -J_{2xz} & -J_{2yz} & J_{2zz} \end{bmatrix}\begin{Bmatrix} -\dot{\theta}_4\,\mathrm{s}\alpha_3\,\mathrm{s}\theta_3 \\ \dot{\theta}_4\,\mathrm{s}\alpha_3\,\mathrm{c}\theta_3 \\ -\dot{\theta}_3 + \dot{\theta}_4\,\mathrm{c}\alpha_3 \end{Bmatrix} \end{aligned}\right)
\end{aligned}
\tag{784}
$$

$$
\begin{gathered}
{}^3\hat{\mathrm{H}}_2 \\
= \\
{}^3\vec{\mathrm{P}}_2 + \varepsilon\, {}^3\vec{\mathrm{H}}_2 \\
= \\
\left\{\begin{matrix}
-\dot{\theta}_3 S_{2y} - \dot{\theta}_4\left[m_2 a_3\, \mathrm{c}\alpha_3\, \mathrm{s}\theta_3 + \left(m_2 s_3 - S_{2z}\right)\mathrm{s}\alpha_3\, \mathrm{c}\theta_3 - S_{2y}\, \mathrm{c}\alpha_3\right] - m_2 \dot{s}_4\, \mathrm{s}\alpha_3\, \mathrm{s}\theta_3 \\
-\dot{\theta}_3 S_{2x} + \dot{\theta}_4\left[m_2 a_3\, \mathrm{c}\alpha_3\, \mathrm{c}\theta_3 - \left(m_2 s_3 - S_{2z}\right)\mathrm{s}\alpha_3\, \mathrm{s}\theta_3 - S_{2x}\, \mathrm{c}\alpha_3\right] + m_2 \dot{s}_4\, \mathrm{s}\alpha_3\, \mathrm{c}\theta_3 \\
\dot{\theta}_4\left(S_{2y}\, \mathrm{s}\theta_3 + S_{2x}\, \mathrm{c}\theta - m_2 a_3\right)\mathrm{s}\alpha_3 - m_2 \dot{s}_3 + m_2 \dot{s}_4\, \mathrm{c}\alpha_3
\end{matrix}\right\} \\
+ \\
\varepsilon\left\{\begin{matrix}
+\dot{\theta}_3 J_{2xz} - \dot{\theta}_4\left[\left(J_{2xx} - S_{2z} s_3\right)\mathrm{s}\alpha_3\, \mathrm{s}\theta_3 + \left(J_{2xy}\, \mathrm{s}\alpha_3 + S_{2z} a_3\, \mathrm{c}\alpha_3\right)\mathrm{c}\theta_3 - S_{2y} a_3\, \mathrm{s}\alpha_3 + J_{2xz}\, \mathrm{c}\alpha_3\right] \\
+ \dot{s}_3 S_{2y} + \dot{s}_4\left(S_{2y}\, \mathrm{c}\alpha_3 - \mathrm{s}\alpha_3\, \mathrm{c}\theta_3\right) \\
\\
+\dot{\theta}_3 J_{2yz} - \dot{\theta}_4\left[\left(J_{2xy}\, \mathrm{s}\alpha_3 + S_{2z} a_3\, \mathrm{c}\alpha_3\right)\mathrm{s}\theta_3 - \left(J_{2yy} - S_{2z} s_3\right)\mathrm{s}\alpha_3\, \mathrm{c}\theta_3 - S_{2x} a_3\, \mathrm{s}\alpha_3 + J_{2yz}\, \mathrm{c}\alpha_3\right] \\
+ \dot{s}_3 S_{2x} - \dot{s}_4\left(S_{2x}\, \mathrm{c}\alpha_3 - S_{2z}\, \mathrm{s}\alpha_3\, \mathrm{s}\theta_3\right) \\
\\
-\dot{\theta}_3 J_{2zz} + \dot{\theta}_4\left\{\left[\left(J_{2xz} + S_{2x} s_3\right)\mathrm{s}\alpha_3 + S_{2y} a_3\, \mathrm{c}\alpha_3\right]\mathrm{s}\theta_3 - \left[\left(J_{2yz} - S_{2y} s_3\right)\mathrm{s}\alpha_3 - S_{2x} a_3\, \mathrm{c}\alpha_3\right]\mathrm{c}\theta_3 - J_{2zz}\, \mathrm{c}\alpha_3\right\} \\
+ \dot{s}_4\left(S_{2y}\, \mathrm{s}\theta_3 + S_{2x}\, \mathrm{c}\theta_3\right)\mathrm{s}\alpha_3
\end{matrix}\right\}
\end{gathered}
\tag{785}
$$

and the dual momentum of the output crank is

$$
{}^4\hat{\mathrm{H}}_3 = m_3 \vec{\mathrm{v}}_3 - \left[{}^4 S_3^4\right]\vec{\omega}_3 + \varepsilon\left(\left[{}^4 S_3^4\right]\vec{\mathrm{v}}_3 + \left[{}^4 J_3^4\right]\vec{\omega}_3\right) \tag{786}
$$

$$
\begin{gathered}
= \\
m_3\begin{Bmatrix} 0 \\ 0 \\ -\dot{s}_4 \end{Bmatrix} - \begin{bmatrix} 0 & -S_{3z} & S_{3y} \\ S_{3z} & 0 & -S_{3x} \\ -S_{3y} & S_{3x} & 0 \end{bmatrix}\begin{Bmatrix} 0 \\ 0 \\ -\dot{\theta}_4 \end{Bmatrix} \\
+\varepsilon\left(\begin{bmatrix} 0 & -S_{3z} & S_{3y} \\ S_{3z} & 0 & -S_{3x} \\ -S_{3y} & S_{3x} & 0 \end{bmatrix}\begin{Bmatrix} 0 \\ 0 \\ -\dot{s}_4 \end{Bmatrix} + \begin{bmatrix} J_{3xx} & -J_{3xy} & -J_{3xz} \\ -J_{3xy} & J_{3yy} & -J_{3yz} \\ -J_{3xz} & -J_{3yz} & J_{3zz} \end{bmatrix}\begin{Bmatrix} 0 \\ 0 \\ -\dot{\theta}_4 \end{Bmatrix}\right)
\end{gathered}
\tag{787}
$$

$$
{}^4\hat{\mathrm{H}}_3 = {}^4\vec{\mathrm{P}}_3 + \varepsilon\, {}^4\vec{\mathrm{H}}_3 = \begin{Bmatrix} -\dot{\theta}_4 S_{3y} \\ \dot{\theta}_4 S_{3x} \\ -m_3 \dot{s}_4 \end{Bmatrix} + \varepsilon\begin{Bmatrix} -\dot{s}_4 S_{3y} + \dot{\theta}_4 J_{3xz} \\ \dot{s}_4 S_{3x} + \dot{\theta}_4 J_{3yz} \\ -\dot{\theta}_4 J_{3zz} \end{Bmatrix} \tag{788}
$$

The time derivatives of the momenta are

$$
\begin{aligned}
&{}^{2}\dot{\hat{\mathrm{H}}}_{1} \\
&= \\
&{}^{2}\dot{\vec{\mathrm{P}}}_{1} + \varepsilon\,{}^{2}\dot{\vec{\mathrm{H}}}_{1} \\
&= \\
&\left\{\begin{matrix}
\ddot{\theta}_1\left(S_{1z}\,\mathrm{s}\alpha_1 - S_{1y}\,\mathrm{c}\alpha_1\right) \\
\ddot{\theta}_1\left(m_1 a_1 + S_{1x}\right)\mathrm{c}\alpha_1 + m_1\ddot{s}_1\,\mathrm{s}\alpha_1 \\
-\ddot{\theta}_1\left(m_1 a_1 + S_{1x}\right)\mathrm{s}\alpha_1 + m_1\ddot{s}_1\,\mathrm{c}\alpha_1
\end{matrix}\right\} \\
&+ \\
&\varepsilon\left\{\begin{matrix}
-\ddot{\theta}_1\left[a_1\left(S_{1z}\,\mathrm{c}\alpha_1 + S_{1y}\,\mathrm{s}\alpha_1\right) + J_{1xy} + J_{1xz}\right] + \ddot{s}_1\left(S_{1y}\,\mathrm{c}\alpha_1 - S_{1z}\,\mathrm{s}\alpha_1\right) \\
-\ddot{\theta}_1\left(S_{1x} a_1\,\mathrm{s}\alpha_1 - J_{1yy} + J_{1yz}\right) - \ddot{s}_1 S_{1x}\,\mathrm{c}\alpha_1 \\
\ddot{\theta}_1\left(S_{1x} a_1\,\mathrm{c}\alpha_1 - J_{1yz} + J_{1zz}\right) + \ddot{s}_1 S_{1x}\,\mathrm{s}\alpha_1
\end{matrix}\right\}
\end{aligned}
\tag{789}
$$

$$
\begin{aligned}
&{}^{3}\dot{\hat{\mathrm{H}}}_{2} \\
&= \\
&{}^{3}\dot{\vec{\mathrm{P}}}_{2} + \varepsilon\,{}^{3}\dot{\vec{\mathrm{H}}}_{2} \\
&= \\
&\left\{\begin{matrix}
-\ddot{\theta}_3 S_{2y} - \ddot{\theta}_4\left[m_2 a_3\,\mathrm{c}\alpha_3\,\mathrm{s}\theta_3 - \left(m_2 s_3 - S_{2z}\right)\mathrm{s}\alpha_3\,\mathrm{c}\theta_3 - S_{2x}\,\mathrm{c}\alpha_3\right] \\
-\dot{\theta}_3\dot{\theta}_4\left[m_2 a_3\,\mathrm{c}\alpha_3\,\mathrm{c}\theta_3 - \left(m_2 s_3 - S_{2z}\right)\mathrm{s}\alpha_3\,\mathrm{s}\theta_3\right] - m_2\ddot{s}_4\,\mathrm{s}\alpha_3\,\mathrm{s}\theta_3 - m_2\dot{\theta}_3\dot{s}_4\,\mathrm{s}\alpha_3\,\mathrm{c}\theta_3 \\
\\
-\ddot{\theta}_3 S_{2x} - \ddot{\theta}_4\left[m_2 a_3\,\mathrm{c}\alpha_3\,\mathrm{c}\theta_3 - \left(m_2 s_3 + S_{2x}\right)\mathrm{s}\alpha_3\,\mathrm{s}\theta_3 - S_{2x}\,\mathrm{c}\alpha_3\right] \\
-\dot{\theta}_3\dot{\theta}_4\left[m_2 a_3\,\mathrm{c}\alpha_3\,\mathrm{s}\theta_3 + \left(m_2 s_3 + S_{2x}\right)\mathrm{s}\alpha_3\,\mathrm{c}\theta_3\right] + m_2\ddot{s}_4\,\mathrm{s}\alpha_3\,\mathrm{c}\theta_3 - m_2\dot{\theta}_3\dot{s}_4\,\mathrm{s}\alpha_3\,\mathrm{s}\theta_3 \\
\\
\ddot{\theta}_4\left(S_{2y}\,\mathrm{s}\theta_3 + S_{2x}\,\mathrm{c}\theta_3 - m_2 a_3\right)\mathrm{s}\alpha_3 + \dot{\theta}_3\dot{\theta}_4\left(S_{2y}\,\mathrm{c}\theta_3 - S_{2x}\,\mathrm{s}\theta_3\right) - m_2\ddot{s}_3 - m_2\ddot{s}_4\,\mathrm{c}\alpha_4
\end{matrix}\right\} \\
&+ \\
&\varepsilon\left\{\begin{matrix}
\ddot{\theta}_3 J_{2xz} - \ddot{\theta}_4\left[\left(J_{2xx} - S_{2z}s_3\right)\mathrm{s}\alpha_3\,\mathrm{s}\theta_3 + \left(J_{2xy}\,\mathrm{s}\alpha_3 + S_{2z}a_3\,\mathrm{c}\alpha_3\right)\mathrm{c}\theta_3 - S_{2y}a_3\,\mathrm{s}\alpha_3 + J_{2xz}\,\mathrm{c}\alpha_3\right] \\
-\dot{\theta}_3\dot{\theta}_4\left[\left(J_{2xx} - S_{2z}s_3\right)\mathrm{s}\alpha_3\,\mathrm{c}\theta_3 - \left(J_{2xy}\,\mathrm{s}\alpha_3 + S_{2z}a_3\,\mathrm{c}\alpha_3\right)\mathrm{s}\theta_3\right] \\
-\ddot{s}_3 S_{2y} + \ddot{s}_4\left(S_{2y}\,\mathrm{c}\alpha_3 - S_{2z}\,\mathrm{s}\alpha_3\,\mathrm{c}\theta_3\right) + \dot{\theta}_3\dot{\theta}_4 S_{2z}\,\mathrm{s}\alpha_3\,\mathrm{s}\theta_3 \\
\\
\ddot{\theta}_3 J_{2yz} - \ddot{\theta}_4\left[\left(J_{2xy}\,\mathrm{s}\alpha_3 + S_{2z}a_3\,\mathrm{c}\alpha_3\right)\mathrm{s}\theta_3 - \left(J_{2yy} - S_{2z}s_3\right)\mathrm{s}\alpha_3\,\mathrm{c}\theta_3 - S_{2x}a_3\,\mathrm{s}\alpha_3 + J_{2yz}\,\mathrm{c}\alpha_3\right] \\
-\dot{\theta}_3\dot{\theta}_4\left[\left(J_{2xy}\,\mathrm{s}\alpha_3 + S_{2z}a_3\,\mathrm{c}\alpha_3\right)\mathrm{c}\theta_3 + \left(J_{2yy} - S_{2z}s_3\right)\mathrm{s}\alpha_3\,\mathrm{s}\theta_3\right] \\
-\ddot{s}_3 S_{2x} - \ddot{s}_4\left(S_{2x}\,\mathrm{c}\alpha_3 + S_{2z}\,\mathrm{s}\alpha_3\,\mathrm{s}\theta_3\right) - \dot{\theta}_3\dot{\theta}_4 S_{2z}\,\mathrm{s}\alpha_3\,\mathrm{c}\theta_3 \\
\\
-\ddot{\theta}_3 J_{2xz} + \ddot{\theta}_4\left\{\left[\left(J_{2xz} - S_{2x}s_3\right)\mathrm{s}\alpha_3 + S_{2y}a_3\,\mathrm{c}\alpha_3\right]\mathrm{s}\theta_3 - \left[\left(J_{2yz} - S_{2y}s_3\right)\mathrm{s}\alpha_3 - S_{2x}a_3\,\mathrm{c}\alpha_3\right]\mathrm{c}\theta_3 + J_{2zz}\,\mathrm{c}\alpha_3\right\} \\
+\dot{\theta}_3\dot{\theta}_4\left\{\left[\left(J_{2xz} - S_{2x}s_3\right)\mathrm{s}\alpha_3 + S_{2y}a_3\,\mathrm{c}\alpha_3\right]\mathrm{c}\theta_3 + \left[\left(J_{2yz} - S_{2y}s_3\right)\mathrm{s}\alpha_3 - S_{2x}a_3\,\mathrm{c}\alpha_3\right]\mathrm{s}\theta_3\right\} \\
+\ddot{s}_4\left(S_{2y}\,\mathrm{s}\theta_3 + S_{2x}\,\mathrm{c}\theta_3\right)\mathrm{s}\alpha_3 + \dot{\theta}_3\dot{s}_4\left(S_{2y}\,\mathrm{c}\theta_3 - S_{2x}\,\mathrm{s}\theta_3\right)\mathrm{s}\alpha_3
\end{matrix}\right\}
\end{aligned}
\tag{790}
$$

$$
{}^{4}\dot{\hat{\mathrm{H}}}_{3} = {}^{4}\dot{\vec{\mathrm{P}}}_{3} + \varepsilon\,{}^{4}\dot{\vec{\mathrm{H}}}_{3} = \left\{\begin{matrix} -\ddot{\theta}_4 S_{3y} \\ \ddot{\theta}_4 S_{3x} \\ -m_3\ddot{s}_4 \end{matrix}\right\} + \varepsilon\left\{\begin{matrix} -\ddot{s}_4 S_{3y} + \ddot{\theta}_4 J_{3xz} \\ \ddot{s}_4 S_{3x} + \ddot{\theta}_4 J_{3yz} \\ -\ddot{\theta}_4 J_{3zz} \end{matrix}\right\} \tag{791}
$$

The inertia forces of each moving link n is obtained by substitution into the dual Euler equation

$$\begin{aligned} &{}^{n+1}\hat{\mathrm{f}}_n \\ &= \\ &\begin{Bmatrix} f_{ni} + \varepsilon t_{ni} \\ f_{nj} + \varepsilon t_{nj} \\ f_{nk} + \varepsilon t_{nk} \end{Bmatrix} \\ &= \\ &\begin{Bmatrix} \dot{P}_{nx} - P_{ny}\omega_{nz} + P_{nz}\omega_{ny} \\ \dot{P}_{ny} - P_{nx}\omega_{nz} + P_{nz}\omega_{nx} \\ \dot{P}_{nx} - P_{nx}\omega_{ny} + P_{ny}\omega_{nx} \end{Bmatrix} + \varepsilon \begin{Bmatrix} \dot{H}_{nx} - H_{ny}\omega_{nz} + H_{nz}\omega_{ny} + P_{ny}v_{nz} + P_{nz}v_{ny} \\ \dot{H}_{ny} + H_{nx}\omega_{nz} - H_{nz}\omega_{nx} + P_{nx}v_{nz} + P_{nz}v_{nx} \\ \dot{H}_{nx} - H_{nx}\omega_{ny} + H_{ny}\omega_{nx} - P_{nx}v_{ny} + P_{ny}v_{nx} \end{Bmatrix} \end{aligned} \tag{792}$$

of the components of the dual vectors given in eqs.(771,782,789) for the input crank, eqs.(772,785,790) for the coupler and eqs.(773,788,791) for the output crank.

Having the inertia forces, the D'Alembert's principle is applied to write the equations of dynamic equilibrium for each of the moving links,

$${}^2_1\hat{\mathrm{M}}\,{}^1\hat{\mathrm{F}}_1 - {}^2\hat{\mathrm{F}}_2 = {}^2\hat{\mathrm{f}}_1 \tag{793}$$

$${}^3_2\hat{\mathrm{M}}\,{}^2\hat{\mathrm{F}}_2 - {}^3\hat{\mathrm{F}}_3 = {}^3\hat{\mathrm{f}}_2 \tag{794}$$

$${}^4_3\hat{\mathrm{M}}\,{}^3\hat{\mathrm{F}}_3 - {}^4\hat{\mathrm{F}}_4 = {}^4\hat{\mathrm{f}}_3 \tag{795}$$

Eqs.(793,794,795) form a simultaneous set of equations which when expanded can be divided into two subsets, one of nine equations in the nine force-component unknowns and the other nine equations in both the nine force-component unknowns and the nine torque-component unknowns. The subset written for the force components can be solved first and the force components which have been obtained can be used as constants in the other subset which can then be solved for the torque components.

Conveniently, both subsets have the same coefficient matrix, so they have the form

$$\begin{bmatrix} c\theta_1 & s\theta_1 & -1 & 0 & 0 & 0 & 0 & 0 & 0 \\ -c\alpha_1 s\theta_1 & c\alpha_1 c\theta_1 & 0 & -1 & 0 & 0 & 0 & 0 & 0 \\ s\alpha_1 s\theta_1 & -s\alpha_1 c\theta_1 & 0 & 0 & 0 & 0 & 0 & 0 & 0 \\ 0 & 0 & c\theta_2 & s\theta_2 & -1 & 0 & 0 & 0 & 0 \\ 0 & 0 & -c\alpha_2 s\theta_2 & c\alpha_2 c\theta_2 & 0 & -1 & 0 & 0 & 0 \\ 0 & 0 & s\alpha_2 s\theta_2 & -s\alpha_2 c\theta_2 & 0 & 0 & 0 & 0 & 0 \\ 0 & 0 & 0 & 0 & c\theta_3 & s\theta_3 & -1 & 0 & 0 \\ 0 & 0 & 0 & 0 & -c\alpha_3 s\theta_3 & c\alpha_3 c\theta_3 & 0 & -1 & 0 \\ 0 & 0 & 0 & 0 & s\alpha_3 s\theta_3 & -s\alpha_3 c\theta_3 & 0 & 0 & -1 \end{bmatrix} \begin{Bmatrix} U_1 \\ U_2 \\ U_3 \\ U_{41} \\ U_5 \\ U_6 \\ U_7 \\ U_8 \\ U_9 \end{Bmatrix} = \begin{Bmatrix} k_1 \\ k_2 \\ k_3 \\ k_4 \\ k_5 \\ k_6 \\ k_7 \\ k_8 \\ k_9 \end{Bmatrix} \quad (796)$$

where for the first subset

$$\begin{Bmatrix} U_1 \\ U_2 \\ U_3 \\ U_4 \\ U_5 \\ U_6 \\ U_7 \\ U_8 \\ U_9 \end{Bmatrix} = \begin{Bmatrix} {}^1F_{1i} \\ {}^1F_{1j} \\ {}^2F_{2i} \\ {}^2F_{2j} \\ {}^3F_{3i} \\ {}^3F_{3j} \\ {}^4F_{4i} \\ {}^4F_{4j} \\ {}^4F_{4k} \end{Bmatrix} \quad (797)$$

$$\begin{Bmatrix} k_1 \\ k_2 \\ k_3 \\ k_4 \\ k_5 \\ k_6 \\ k_7 \\ k_8 \\ k_9 \end{Bmatrix} = \begin{Bmatrix} {}^2f_{1i} \\ {}^2f_{1j} - {}^1F_{1k}\, s\alpha_1 \\ {}^2f_{1k} - {}^1F_{1k}\, c\alpha_1 + {}^2F_{2k} \\ {}^3f_{2i} \\ {}^3f_{2j} - {}^2F_{2k}\, s\alpha_2 \\ {}^2f_{2k} - {}^2F_{2k}\, c\alpha_2 + {}^3F_{3k} \\ {}^4f_{3i} \\ {}^4f_{3j} - {}^3F_{3k}\, s\alpha_3 \\ {}^4f_{3k} - {}^3F_{3k}\, c\alpha_3 \end{Bmatrix} \quad (798)$$

and for the other subset

$$\begin{Bmatrix} U_1 \\ U_2 \\ U_3 \\ U_4 \\ U_5 \\ U_6 \\ U_7 \\ U_8 \\ U_9 \end{Bmatrix} = \begin{Bmatrix} {}^1T_{1i} \\ {}^1T_{1j} \\ {}^2T_{2i} \\ {}^2T_{2j} \\ {}^3T_{3i} \\ {}^3T_{3j} \\ {}^4T_{4i} \\ {}^4T_{4j} \\ {}^4T_{4k} \end{Bmatrix} \tag{799}$$

$$\begin{Bmatrix} k_1 \\ k_2 \\ k_3 \\ k_4 \\ k_5 \\ k_6 \\ k_7 \\ k_8 \\ k_9 \end{Bmatrix} = \begin{Bmatrix} {}^2t_{1i} + s_1\left({}^1F_{1i}\,s\theta_1 - {}^1F_{1j}\,c\theta_1\right) \\ {}^2t_{1j} - {}^1T_{1k}\,s\alpha_1 - a_1\left[s\alpha_1\left({}^1F_{1i}\,s\theta_1 - {}^1F_{1j}\,c\theta_1\right) + {}^1F_{1k}\,c\alpha_1\right] + s_1\,c\alpha_1\left({}^1F_{1i}\,c\theta_1 + {}^1F_{1j}\,s\theta_1\right) \\ {}^2t_{1j} - {}^1T_{1k}\,c\alpha_1 - a_1\left[c\alpha_1\left({}^1F_{1i}\,s\theta_1 - {}^1F_{1j}\,c\theta_1\right) + {}^1F_{1k}\,s\alpha_1\right] - s_1\,s\alpha_1\left({}^1F_{1i}\,c\theta_1 + {}^1F_{1j}\,s\theta_1\right) \\ {}^3t_{2i} + s_2\left({}^2F_{2i}\,s\theta_2 - {}^2F_{2j}\,c\theta_2\right) \\ {}^3t_{2j} - a_2\left[s\alpha_2\left({}^2F_{2i}\,s\theta_2 - {}^2F_{2j}\,c\theta_2\right) + {}^2F_{2k}\,c\alpha_2\right] + s_2\,c\alpha_2\left({}^2F_{2i}\,c\theta_2 + {}^2F_{2j}\,s\theta_2\right) \\ {}^3t_{2j} - a_2\left[c\alpha_2\left({}^2F_{2i}\,s\theta_2 - {}^2F_{2j}\,c\theta_2\right) + {}^2F_{2k}\,s\alpha_2\right] - s_2\,s\alpha_2\left({}^2F_{2i}\,c\theta_2 + {}^2F_{2j}\,s\theta_2\right) \\ {}^4t_{13} + s_3\left({}^3F_{3i}\,s\theta_3 - {}^3F_{3j}\,c\theta_3\right) \\ {}^4t_{3j} - a_3\left[s\alpha_3\left({}^3F_{3i}\,s\theta_3 - {}^3F_{3j}\,c\theta_3\right) + {}^3F_{3k}\,c\alpha_3\right] + s_3\,c\alpha_3\left({}^3F_{3i}\,c\theta_3 + {}^3F_{3j}\,s\theta_3\right) \\ {}^4t_{3j} - a_3\left[c\alpha_3\left({}^3F_{3i}\,s\theta_3 - {}^3F_{3j}\,c\theta_3\right) + {}^3F_{3k}\,s\alpha_3\right] - s_3\,s\alpha_3\left({}^3F_{3i}\,F\theta_3 + {}^3F_{3j}\,s\theta_3\right) \end{Bmatrix} \tag{800}$$

The solutions of eq.(796) are

$$U_1 = \frac{\begin{bmatrix} k_1\,s\alpha_1\,s\alpha_2\,c\theta_1\,s\theta_2 \\ -k_2\,s\alpha_1\,s\alpha_2\,c\theta_1\,c\theta_2 \\ +k_3\,s\alpha_2\left(s\theta_1\,s\theta_2 - c\alpha_1\,c\theta_1\,c\theta_2\right) \\ +k_6\,s\alpha_1\,c\theta_1 \end{bmatrix}}{s\alpha_1\,s\alpha_2\,s\theta_2} \tag{801}$$

$$U_2 = \frac{\begin{bmatrix} k_1\,s\alpha_1\,s\alpha_2\,s\theta_1\,s\theta_2 \\ -k_2\,s\alpha_1\,s\alpha_2\,s\theta_1\,c\theta_2 \\ -k_3\,s\alpha_2\left(c\theta_1\,s\theta_2 + c\alpha_1\,s\theta_1\,c\theta_2\right) \\ +k_6\,s\alpha_1\,c\theta_1 \end{bmatrix}}{s\alpha_1\,s\alpha_2\,s\theta_2} \tag{802}$$

$$U_3 = \frac{-k_2\,s\alpha_1\,s\alpha_2\,c\theta_2 - k_3\,c\alpha_1\,s\alpha_2\,c\theta_2 + k_6\,s\alpha_1}{s\alpha_1\,s\alpha_2\,s\theta_2} \tag{803}$$

$$U_4 = -\frac{k_3\,\mathrm{c}\alpha_1}{\mathrm{s}\alpha_1} - k_2 \tag{804}$$

$$U_5 = \frac{-k_2\,\mathrm{s}\alpha_1\,\mathrm{s}\alpha_2 - k_3\,\mathrm{c}\alpha_1\,\mathrm{s}\alpha_2 - k_4\,\mathrm{s}\alpha_1\,\mathrm{s}\alpha_2\,\mathrm{s}\theta_2 - k_6\,\mathrm{s}\alpha_1\,\mathrm{c}\theta_2}{\mathrm{s}\alpha_1\,\mathrm{s}\alpha_2\,\mathrm{s}\theta_2} \tag{805}$$

$$U_6 = -\frac{k_6\,\mathrm{c}\alpha_2}{\mathrm{s}\alpha_2} - k_5 \tag{806}$$

$$U_7 = \frac{\left[\begin{array}{c} -k_2\,\mathrm{s}\alpha_1\,\mathrm{s}\alpha_2\,\mathrm{c}\theta_3 \\ -k_3\,\mathrm{c}\alpha_1\,\mathrm{s}\alpha_2\,\mathrm{c}\theta_3 \\ -k_4\,\mathrm{s}\alpha_1\,\mathrm{s}\alpha_2\,\mathrm{s}\theta_2\,\mathrm{c}\theta_3 \\ -k_5\,\mathrm{s}\alpha_1\,\mathrm{s}\alpha_2\,\mathrm{s}\theta_2\,\mathrm{s}\theta_3 \\ +k_6\,\mathrm{s}\alpha_1(\mathrm{c}\theta_2\,\mathrm{c}\theta_3 - \mathrm{c}\alpha_2\,\mathrm{s}\theta_2\,\mathrm{s}\theta_3) \\ -k_7\,\mathrm{s}\alpha_1\,\mathrm{s}\alpha_2\,\mathrm{s}\theta_2 \end{array}\right]}{\mathrm{s}\alpha_1\,\mathrm{s}\alpha_2\,\mathrm{s}\theta_2} \tag{807}$$

$$U_8 = \frac{\left[\begin{array}{c} k_2\,\mathrm{s}\alpha_1\,\mathrm{s}\alpha_2\,\mathrm{c}\alpha_3\,\mathrm{s}\theta_3 \\ +k_3\,\mathrm{c}\alpha_1\,\mathrm{s}\alpha_2\,\mathrm{c}\alpha_3\,\mathrm{s}\theta_3 \\ +k_4\,\mathrm{s}\alpha_1\,\mathrm{s}\alpha_2\,\mathrm{c}\alpha_3\,\mathrm{s}\theta_2\,\mathrm{s}\theta_3 \\ -k_5\,\mathrm{s}\alpha_1\,\mathrm{s}\alpha_2\,\mathrm{c}\alpha_3\,\mathrm{s}\theta_2\,\mathrm{c}\theta_3 \\ -k_6\,\mathrm{s}\alpha_1\,\mathrm{c}\alpha_3(\mathrm{c}\theta_2\,\mathrm{s}\theta_3 + \mathrm{c}\alpha_2\,\mathrm{s}\theta_2\,\mathrm{c}\theta_3) \\ -k_8\,\mathrm{s}\alpha_1\,\mathrm{s}\alpha_2\,\mathrm{s}\theta_2 \end{array}\right]}{\mathrm{s}\alpha_1\,\mathrm{s}\alpha_2\,\mathrm{s}\theta_2} \tag{808}$$

$$U_8 = \frac{\left[\begin{array}{c} -k_2\,\mathrm{s}\alpha_1\,\mathrm{s}\alpha_2\,\mathrm{s}\alpha_3\,\mathrm{s}\theta_3 \\ -k_3\,\mathrm{c}\alpha_1\,\mathrm{s}\alpha_2\,\mathrm{s}\alpha_3\,\mathrm{s}\theta_3 \\ -k_4\,\mathrm{s}\alpha_1\,\mathrm{s}\alpha_2\,\mathrm{s}\alpha_3\,\mathrm{s}\theta_2\,\mathrm{s}\theta_3 \\ +k_5\,\mathrm{s}\alpha_1\,\mathrm{s}\alpha_2\,\mathrm{s}\theta_2\,\mathrm{c}\theta_3 \\ +k_6\,\mathrm{s}\alpha_1\,\mathrm{s}\alpha_3(\mathrm{c}\theta_2\,\mathrm{s}\theta_3 + \mathrm{c}\alpha_2\,\mathrm{s}\theta_2\,\mathrm{c}\theta_3) \\ -k_9\,\mathrm{s}\alpha_1\,\mathrm{s}\alpha_2\,\mathrm{s}\theta_2 \end{array}\right]}{\mathrm{s}\alpha_1\,\mathrm{s}\alpha_2\,\mathrm{s}\theta_2} \tag{809}$$

The solution for the force components can be obtained by substitution of eq.(798) into eqs.(801-809). The force components are used in evaluating eq.(800), the results of which are then substituted into eqs.(801-809) to obtain the torque components.

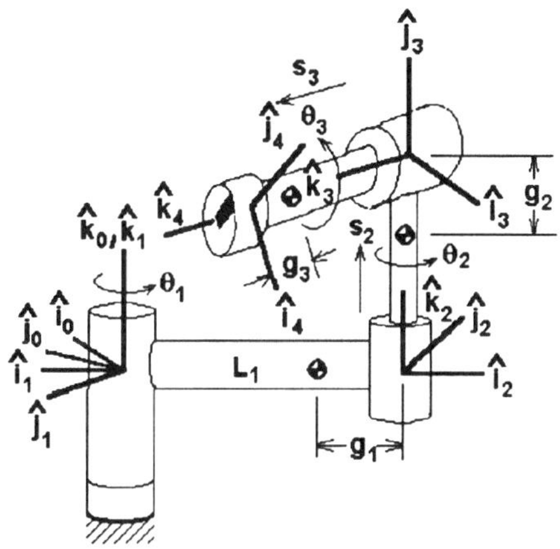

Figure 63 Robot.

Exercises - Dynamic Analysis

1. Consider the robot shown in Figure 63. The joint-link parameter table is

i	θ_i	s_i	α_i	a_i
1	θ_1	0	0	L_1
2	θ_2	s_2	$\pi/2$	0
3	θ_3	s_3	0	0

Various velocities are

$$ {}^2\hat{V}_{20}^2 = \begin{Bmatrix} 0 \\ \varepsilon L_1\dot{\theta}_1 \\ \dot{\theta}_1 + \dot{\theta}_2 + \varepsilon\dot{s}_2 \end{Bmatrix} $$

$$ {}^3\hat{V}_{30}^3 = \begin{Bmatrix} \varepsilon L_1\, s\theta_2\dot{\theta}_1 \\ \dot{\theta}_1 + \dot{\theta}_2 + \varepsilon\dot{s}_2 \\ \dot{\theta}_3 + \varepsilon\left(\dot{s}_3 - L_1\, c\theta_2\dot{\theta}_1\right) \end{Bmatrix} $$

$$ {}^4\hat{V}_{30}^4 = \begin{Bmatrix} (\dot{\theta}_1 + \dot{\theta}_2)s\theta_3 + \varepsilon\left[L_1 s\theta_2\, c\theta_3\dot{\theta}_1 + s\theta_3\dot{s}_2 + s_3\, c\theta_3(\dot{\theta}_1 + \dot{\theta}_2)\right] \\ (\dot{\theta}_1 + \dot{\theta}_2)c\theta_3 + \varepsilon\left[-L_1\, s\theta_2\, s\theta_3\dot{\theta}_1 + c\theta_3\dot{s}_2 - s_3\, s\theta_3(\dot{\theta}_1 + \dot{\theta}_2)\right] \\ \dot{\theta}_3 + \varepsilon(\dot{s}_3 - L_1\, c\theta_2\dot{\theta}_1) \end{Bmatrix} $$

The mass of the links is distributed such that

$$^{2}\vec{G}_1 = \begin{Bmatrix} g_1 \\ 0 \\ 0 \end{Bmatrix} \quad ^{3}\vec{G}_2 = \begin{Bmatrix} 0 \\ g_2 \\ 0 \end{Bmatrix} \quad ^{4}\vec{G}_3 = \begin{Bmatrix} 0 \\ 0 \\ g_3 \end{Bmatrix}$$

$$\left[{}^{2}J_1\right] = m_1 \begin{bmatrix} k_{1x}^2 & 0 & 0 \\ 0 & k_{1y}^2 & 0 \\ 0 & 0 & k_{1z}^2 \end{bmatrix} \quad \left[{}^{3}J_2\right] = m_2 \begin{bmatrix} k_{2x}^2 & 0 & 0 \\ 0 & k_{2y}^2 & 0 \\ 0 & 0 & k_{2z}^2 \end{bmatrix} \quad \left[{}^{4}J_3\right] = m_3 \begin{bmatrix} k_{3x}^2 & 0 & 0 \\ 0 & k_{3y}^2 & 0 \\ 0 & 0 & k_{3z}^2 \end{bmatrix}$$

Evaluate the inertia dual forces of the moving links. Write the equations of dynamic equilibrium.

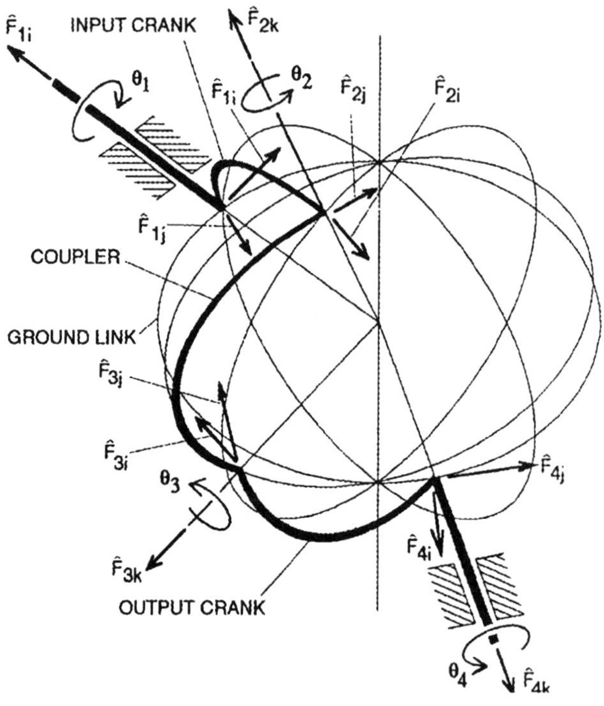

Figure 64 Cardan joint.

2. The Cardan joint is a spherical four-bar mechanism whose ground link has some twist α_4 and whose moving links have twists of 90°. Such a mechanism is illustrated in Figure 64. Write the equations of dynamic equilibrium for the Cardan joint and solve for the internal force and torque components and the output torque component.

Chapter 8 - Dual-Number Programming

The Utility of Dual-Number Programming

The advantage of formulating analysis of mechanical systems in terms of duplex quantities is the simplicity which is obtained. The expressions relating dual-angle dimensions, dual-angle displacements, dual velocities, dual momenta and dual forces are relatively simple and easily managed. However, when the resulting expression is obtained, the simplicity is lost when it must be expanded to yield an expression for the dual component. Considering that a multiplication of two duplex numbers requires three scalar multiplications and an addition, or that a division of two duplex numbers involves two multiplications, one subtraction and two scalar divisions, it is readily realized that expansion to obtain the dual component of an expression so that translations, linear velocities and torques can be calculated is a rather arduous business.

The amount of work required to evaluate the dual component of even an expression of moderate or larger size is intimidating and there is ample opportunity for errors to occur. Symbolic-manipulator software is helpful with the labor involved, but it is expensive and requires effort to learn its use. Very often the expressions which are obtained are too large to be useful for interpreting the influences of variables. This is particularly so in dynamics problems where expressions for the inertia loads in mechanisms with several links can occupy a page or more.

These expansions are required when traditional programming languages such as Fortran or Basic are used. This results in a lot of work before programming can begin and a lot of checking and debugging thereafter. What is needed is a way of handling a duplex number as a single entity and calculating the dual components automatically the way that Fortran does with its complex data type. It is not possible to define new data types in Fortran or Basic, but a duplex number can be defined as data type in the C++ programming language and mathematical functions can be overloaded so that duplex-type variables can be used as arguments.

The ability of dual-number programming very much simplifies the analysis of mechanical systems. Besides getting programs running quickly, the chance of errors is greatly reduced when the dual components are evaluated transparent to the programmer.

The Duplex Header File

The object-oriented features of the C++ programming language allow the use of duplex variables in programs compiled by ordinary, inexpensive, readily available commercial compilers. The definition of the duplex data type and the overloading of operators to implement arithmetic operations with duplex numbers and of functions to handle duplex variables as arguments can be contained in a header file which is made up once and then used with any program in which duplex variables are employed.

`Duplex.hpp` is such a header file. It defines the duplex data type, overloads arithmetic operators to handle duplex numbers, overloads trigonometric functions to handle duplex arguments and defines some other useful functions.

The addition, subtraction, multiplication and division operators are overloaded so that they will handle duplex variables or combinations of double and duplex variables.

The `prim` and `dual` functions are defined to return the primary and dual component of a duplex argument respectively.

The `sin`, `cos` and `tan` functions are overloaded to return the trigonometric functions of duplex arguments.

The `deg2rad` and `rad2deg` functions are useful in the computation of mechanical motions. They provide the conversion between radial measure in degrees and in radians of an angle expressed as a double variable or the primary component of an angle expressed as a duplex variable. Function `rad2deg` is written so that the angle returned is in the range $0<360°$. Thus, if the argument was -7 radians, the value returned would be 318.93° rather than $-401.07°$ or $-41.07°$. The programmer can change this if desired.

Another set of conversion functions provided are `rpm2radpersec` and `radpersec2rpm` which convert the angular component of a dual velocity between expression in rpm or in radians per second or vice versa as is necessary.

Function `pow` provides for raising a duplex variable to a power and function `sqrt` provides the square root of a duplex variable.

The `atan` function is overloaded to return the arctangent of a duplex argument. An `atan2` function which will accept two duplex arguments is specified and is analogous to that featured in the Fortran language.

The assignment operators and unary minus sign are overloaded for duplex variables.

Output- and input-stream insertion operators are overloaded to handle duplex variables.

With the details of dual-number operations buried in the header file, source code can be written entirely in duplex variables, only considering the primary and duplex components during the input and output operations and even that is provided for. The header file `duplex.hpp` can be linked with source code to program calculation with dual variables.

RCCC, Spatial-Mechanism Displacement-Analysis Program

The ease with which the displacements, velocities and dynamic reactions of a spatial four-bar mechanism can be calculated using C++ dual-number programming is demonstrated by the program RCCC. The source code `RCCC.cpp` is linked with the header file `duplex.hpp` for compilation.

The program begins with `#include` statements for the header files for mathematical, input-output and dual-number operations.

Two objects are then defined. The first of these is `start_and_step` which is concerned with the initial state of the mechanism's displacements and speeds and the size of the step in input rotation. The other object `linkage` deals with the dimensions of the mechanism, its displacements, joint speeds, absolute velocities, momenta, time derivatives of momenta, inertia forces and reaction forces.

The `main` function of the program begins with the declaration of several variables and then the declaration of `mechanism` as a `linkage` object and `index` as a `start_and_step` object.

The program then displays its name, RCCC, on the screen and requests that the user indicate if the mechanism dimensions will be specified or if default mechanism dimensions are to be used, a convenience when demonstrating the program. In the latter case, the functions `get_links` and `get_start_and_step` are executed in which inquiries are implemented through the screen with input through the keyboard to assign numeric values to the dual angles representing the dimensions of the four links, the mass of the links, the first moments of inertia, second moments of inertia and the products of inertia of the three moving links, the translational displacement and starting angle of the input link, the size of step in the input-link rotation and the angular and linear input speeds. If the default conditions are selected, then values are assigned to these quantities by functions `default_links` and `default_start_and_step`.

The user is then asked through the screen if the calculated data is to be saved in a disk file. If the response is positive, then a file name is requested of the user, the file is opened and column headers are written in the first line of the data file. This last action may have to be omitted if the user's plotting program cannot handle it.

The function calls for the calculations, and the screen and file outputs are contained in a `while` statement in which the input shaft is stepped through its rotation from 0 to 360°.

Using the input-joint displacements as its arguments, the first function call is to the function `displacements` which returns the calculated values of the joint displacements. First the input-joint rotation and translation are combined as a dual angle where the primary component is specified in radians. The factors $\hat{D}$, $\hat{E}$ and $\hat{F}$ are calculated. Without dual-number variables, it is practical to consider only the primary components of these factors to compute the primary component of output-joint displacement dual angle (the rotation), requiring that the dual component (the translation) be calculated separately. Having dual variables, we can directly implement

$$\hat{t} = \frac{-\hat{D} - \sqrt{\hat{D}^2 + \hat{E}^2 - \hat{F}^2}}{\left(\hat{F} - \hat{E}\right)} \tag{810}$$

and then

$$\hat{\theta}_4 = 2\tan^{-1}\hat{t} \tag{811}$$

to quickly obtain both the rotation and translation at the output joint combined as a dual angle.

The displacements at the intermediate joints are readily obtained by computing the dual-number values of their sines and cosines as

$$s\hat{\theta}_2 = \frac{\left(c\hat{\alpha}_3\, s\hat{\alpha}_4 + s\hat{\alpha}_3\, c\hat{\alpha}_4\, c\hat{\theta}_4\right)s\hat{\theta}_1 + s\hat{\alpha}_3\, c\hat{\theta}_1\, s\hat{\theta}_4}{s\hat{\alpha}_2} \tag{812}$$

$$c\hat{\theta}_2 = \frac{c\hat{\alpha}_1\, c\hat{\alpha}_2 - c\hat{\alpha}_3\, c\hat{\alpha}_4 + s\hat{\alpha}_3\, s\hat{\alpha}_4\, c\hat{\theta}_4}{s\hat{\alpha}_1\, s\hat{\alpha}_2} \tag{813}$$

$$s\hat{\theta}_3 = \frac{c\hat{\alpha}_1\, s\hat{\alpha}_4\, s\hat{\theta}_4 + s\hat{\alpha}_1\left(s\hat{\theta}_1\, c\hat{\theta}_4 + c\hat{\alpha}_4\, c\hat{\theta}_1\, s\hat{\theta}_4\right)}{s\hat{\alpha}_2} \tag{814}$$

$$c\hat{\theta}_3 = \frac{-c\hat{\alpha}_1\, c\hat{\alpha}_4 + c\hat{\alpha}_2\, c\hat{\alpha}_3 + s\hat{\alpha}_1\, s\hat{\alpha}_4\, c\hat{\theta}_1}{s\hat{\alpha}_2\, s\hat{\alpha}_3} \tag{815}$$

and then using the double-argument arctangent function to obtain the intermediate displacements as dual angles

$$\hat{\theta}_2 = \text{atan2}\left(\sin\hat{\theta}_2, \cos\hat{\theta}_2\right) \tag{816}$$

$$\hat{\theta}_3 = \text{atan2}\left(\sin\hat{\theta}_3, \cos\hat{\theta}_3\right) \tag{817}$$

The next function call in the `main` function is to the function `velocities` using the angular velocity and the linear velocity of the input shaft as arguments. In the velocities function the input-joint speed is formulated as a duplex variable and then the other three joint speeds are calculated as duplex quantities.

With the joint speeds calculated, the `main` function then calls the function `absolute_velocities`. Using the joint speeds, the link dimensions and the joint displacements, all dual quantities, the three components of each absolute velocity vector of each of the three moving links are calculated as dual quantities.

The `momenta` function is called next. The momenta calculated for the previous input-rotation step is first saved since it will be later used in calculating the derivatives of the momenta. Using dual-number equations with the mass quantities and absolute velocities, the three components of each of the three moving links are calculated.

Using the function `derivatives_of_momenta`, the time derivative of the three components of the momentum vectors of each of the moving links is calculated using a Newton-quotient formulation based on the present value of the momentum component, the value of the momentum component at the previous step of input-shaft rotation and the step size and input-shaft angular velocity. Because nothing has been calculated previously, the values of the time derivatives of the momentum vectors for the initial step will be spurious, affecting the calculation of the inertia dual forces and the reaction dual forces. If this is a problem, then the correct values can be obtained from the results of the last step where the mechanism has returned to its original configuration.

With the calculated values of the dual absolute-velocity vectors, the dual-momentum vectors and the time derivatives of the dual-momentum vectors, the dual inertia-force vectors are calculated in the function `inertia_forces` which is then called from the `main` function.

Next the dual reaction forces are calculated when the main function calls the `reaction_forces` function. It is recalled that the equations of dynamic equilibrium of the mechanism take the form

$$\begin{bmatrix} c\hat{\theta}_1 & s\hat{\theta}_1 & -1 & 0 & 0 & 0 & 0 & 0 & 0 \\ -c\hat{\alpha}_1 s\hat{\theta}_1 & c\hat{\alpha}_1 c\hat{\theta}_1 & 0 & -1 & 0 & 0 & 0 & 0 & 0 \\ s\hat{\alpha}_1 s\hat{\theta}_1 & -s\hat{\alpha}_1 c\hat{\theta}_1 & 0 & 0 & 0 & 0 & 0 & 0 & 0 \\ 0 & 0 & c\hat{\theta}_2 & s\hat{\theta}_2 & -1 & 0 & 0 & 0 & 0 \\ 0 & 0 & -c\hat{\alpha}_2 s\hat{\theta}_2 & c\hat{\alpha}_2 c\hat{\theta}_2 & 0 & -1 & 0 & 0 & 0 \\ 0 & 0 & s\hat{\alpha}_2 s\hat{\theta}_2 & -s\hat{\alpha}_2 c\hat{\theta}_2 & 0 & 0 & 0 & 0 & 0 \\ 0 & 0 & 0 & 0 & c\hat{\theta}_3 & s\hat{\theta}_3 & -1 & 0 & 0 \\ 0 & 0 & 0 & 0 & -c\hat{\alpha}_3 s\hat{\theta}_3 & c\hat{\alpha}_3 c\hat{\theta}_3 & 0 & -1 & 0 \\ 0 & 0 & 0 & 0 & s\hat{\alpha}_3 s\hat{\theta}_3 & -s\hat{\alpha}_3 c\hat{\theta}_3 & 0 & 0 & -1 \end{bmatrix} \begin{Bmatrix} {}^1\hat{F}_{1i} \\ {}^1\hat{F}_{1j} \\ {}^2\hat{F}_{2i} \\ {}^2\hat{F}_{2j} \\ {}^3\hat{F}_{3i} \\ {}^3\hat{F}_{3j} \\ {}^4\hat{F}_{4i} \\ {}^4\hat{F}_{4j} \\ {}^4\hat{F}_{4k} \end{Bmatrix} = \begin{Bmatrix} \hat{k}_1 \\ \hat{k}_2 \\ \hat{k}_3 \\ \hat{k}_4 \\ \hat{k}_5 \\ \hat{k}_6 \\ \hat{k}_7 \\ \hat{k}_8 \\ \hat{k}_9 \end{Bmatrix} \tag{818}$$

where the vector of constants is

$$\begin{Bmatrix} \hat{k}_1 \\ \hat{k}_2 \\ \hat{k}_3 \\ \hat{k}_4 \\ \hat{k}_5 \\ \hat{k}_6 \\ \hat{k}_7 \\ \hat{k}_8 \\ \hat{k}_9 \end{Bmatrix} = \begin{Bmatrix} {}^2\hat{f}_{1i} \\ {}^2\hat{f}_{1j} - {}^1\hat{F}_{1k}\, s\hat{\alpha}_1 \\ {}^2\hat{f}_{1k} - {}^1\hat{F}_{1k}\, c\hat{\alpha}_1 + {}^2\hat{F}_{2k} \\ {}^3\hat{f}_{2i} \\ {}^3\hat{f}_{2j} - {}^2\hat{F}_{2k}\, s\hat{\alpha}_2 \\ {}^3\hat{f}_{2k} - {}^2\hat{F}_{2k}\, c\hat{\alpha}_2 + {}^3\hat{F}_{3k} \\ {}^4\hat{f}_{3i} \\ {}^4\hat{f}_{3j} - {}^3\hat{F}_{3k}\, s\hat{\alpha}_2 \\ {}^4\hat{f}_{3k} - {}^3\hat{F}_{3k}\, c\hat{\alpha}_2 \end{Bmatrix} \tag{819}$$

Without a duplex variable type, this equation has to be expanded and its primary and dual components solved separately. As has been observed, both the primary and dual components of eq.(818) have the same coefficient matrix which does simplify matters, but we can solve the equation set directly in terms of duplex quantities, yielding the dual reaction forces directly. The solutions are

$${}^1\hat{F}_{1i} = \frac{\begin{bmatrix} \hat{k}_1\, s\hat{\alpha}_1\, s\hat{\alpha}_2\, c\hat{\theta}_1\, s\hat{\theta}_2 \\ -\hat{k}_2\, s\hat{\alpha}_1\, s\hat{\alpha}_2\, c\hat{\theta}_1\, c\hat{\theta}_2 \\ +\hat{k}_3\, s\hat{\alpha}_2 \left(s\hat{\theta}_1\, s\hat{\theta}_2 - c\hat{\alpha}_1\, c\hat{\theta}_1\, c\hat{\theta}_2 \right) \\ +\hat{k}_6\, s\hat{\alpha}_1\, c\hat{\theta}_1 \end{bmatrix}}{s\hat{\alpha}_1\, s\hat{\alpha}_2\, s\hat{\theta}_2} \tag{820}$$

$${}^1\hat{F}_{1j} = \frac{\begin{bmatrix} \hat{k}_1\, s\hat{\alpha}_1\, s\hat{\alpha}_2\, s\hat{\theta}_1\, s\hat{\theta}_2 \\ -\hat{k}_2\, s\hat{\alpha}_1\, s\hat{\alpha}_2\, s\hat{\theta}_1\, c\hat{\theta}_2 \\ -\hat{k}_3\, s\hat{\alpha}_2 \left(c\hat{\theta}_1\, s\hat{\theta}_2 + c\hat{\alpha}_1\, s\hat{\theta}_1\, c\hat{\theta}_2 \right) \\ +\hat{k}_6\, s\hat{\alpha}_1\, c\hat{\theta}_1 \end{bmatrix}}{s\hat{\alpha}_1\, s\hat{\alpha}_2\, s\hat{\theta}_2} \tag{821}$$

$$ {}^{2}\hat{F}_{2i} = \frac{-\hat{k}_2\, s\hat{\alpha}_1\, s\hat{\alpha}_2\, c\hat{\theta}_2 - \hat{k}_3\, c\hat{\alpha}_1\, s\hat{\alpha}_2\, c\hat{\theta}_2 + \hat{k}_6\, s\hat{\alpha}_1}{s\hat{\alpha}_1\, s\hat{\alpha}_2\, s\hat{\theta}_2} \tag{822} $$

$$ {}^{2}\hat{F}_{2j} = -\frac{\hat{k}_3\, c\hat{\alpha}_1}{s\hat{\alpha}_1} - \hat{k}_2 \tag{823} $$

$$ {}^{3}\hat{F}_{3i} = \frac{-\hat{k}_2\, s\hat{\alpha}_1\, s\hat{\alpha}_2 - \hat{k}_3\, c\hat{\alpha}_1\, s\hat{\alpha}_2 - \hat{k}_4\, s\hat{\alpha}_1\, s\hat{\alpha}_2\, s\hat{\theta}_2 - \hat{k}_6\, s\hat{\alpha}_1\, c\hat{\theta}_2}{s\hat{\alpha}_1\, s\hat{\alpha}_2\, s\hat{\theta}_2} \tag{824} $$

$$ {}^{3}\hat{F}_{3j} = -\frac{\hat{k}_6\, c\hat{\alpha}_2}{s\hat{\alpha}_2} - \hat{k}_5 \tag{825} $$

$$ {}^{4}\hat{F}_{4i} = \frac{\begin{bmatrix} -\hat{k}_2\, s\hat{\alpha}_1\, s\hat{\alpha}_2\, c\hat{\theta}_3 \\ -\hat{k}_3\, c\hat{\alpha}_1\, s\hat{\alpha}_2\, c\hat{\theta}_3 \\ -\hat{k}_4\, s\hat{\alpha}_1\, s\hat{\alpha}_2\, s\hat{\theta}_2\, c\hat{\theta}_3 \\ -\hat{k}_5\, s\hat{\alpha}_1\, s\hat{\alpha}_2\, s\hat{\theta}_2\, s\hat{\theta}_3 \\ +\hat{k}_6\, s\hat{\alpha}_1\left(c\hat{\theta}_2\, c\hat{\theta}_3 - c\hat{\alpha}_2\, s\hat{\theta}_2\, s\hat{\theta}_3\right) \\ -\hat{k}_7\, s\hat{\alpha}_1\, s\hat{\alpha}_2\, s\hat{\theta}_2 \end{bmatrix}}{s\hat{\alpha}_1\, s\hat{\alpha}_2\, s\hat{\theta}_2} \tag{826} $$

$$ {}^{4}\hat{F}_{4j} = \frac{\begin{bmatrix} \hat{k}_2\, s\hat{\alpha}_1\, s\hat{\alpha}_2\, c\hat{\alpha}_3\, s\hat{\theta}_3 \\ +\hat{k}_3\, c\hat{\alpha}_1\, s\hat{\alpha}_2\, c\hat{\alpha}_3\, s\hat{\theta}_3 \\ +\hat{k}_4\, s\hat{\alpha}_1\, s\hat{\alpha}_2\, c\hat{\alpha}_3\, s\hat{\theta}_2\, s\hat{\theta}_3 \\ -\hat{k}_5\, s\hat{\alpha}_1\, s\hat{\alpha}_2\, c\hat{\alpha}_3\, s\hat{\theta}_2\, c\hat{\theta}_3 \\ -\hat{k}_6\, s\hat{\alpha}_1\, c\hat{\alpha}_3\left(c\hat{\theta}_2\, s\hat{\theta}_3 + c\hat{\alpha}_2\, s\hat{\theta}_2\, c\hat{\theta}_3\right) \\ -\hat{k}_8\, s\hat{\alpha}_1\, s\hat{\alpha}_2\, s\hat{\theta}_2 \end{bmatrix}}{s\hat{\alpha}_1\, s\hat{\alpha}_2\, s\hat{\theta}_2} \tag{827} $$

$$ {}^{4}\hat{F}_{4k} = \frac{\begin{bmatrix} -\hat{k}_2\, s\hat{\alpha}_1\, s\hat{\alpha}_2\, s\hat{\alpha}_3\, s\hat{\theta}_3 \\ -\hat{k}_3\, c\hat{\alpha}_1\, s\hat{\alpha}_2\, s\hat{\alpha}_3\, s\hat{\theta}_3 \\ -\hat{k}_4\, s\hat{\alpha}_1\, s\hat{\alpha}_2\, s\hat{\alpha}_3\, s\hat{\theta}_2\, s\hat{\theta}_3 \\ +\hat{k}_5\, s\hat{\alpha}_1\, s\hat{\alpha}_2\, s\hat{\theta}_2\, c\hat{\theta}_3 \\ +\hat{k}_6\, s\hat{\alpha}_1\, s\hat{\alpha}_3\left(c\hat{\theta}_2\, s\hat{\theta}_3 + c\hat{\alpha}_2\, s\hat{\theta}_2\, c\hat{\theta}_3\right) \\ -\hat{k}_9\, s\hat{\alpha}_1\, s\hat{\alpha}_2\, s\hat{\theta}_2 \end{bmatrix}}{s\hat{\alpha}_1\, s\hat{\alpha}_2\, s\hat{\theta}_2} \tag{828} $$

Using eqs.(819-828), the reaction dual forces can be evaluated directly, and using the `prim` and `dual` functions, the forces and torques can be obtained. The ability to calculate these results without getting involved with the arduous mathematics involved with handling the information contained in the dual components of the dimensions and variables, is a significant advantage obtained from using the dual number approach.

The Ideal Cardan Joint

Sample calculations have been conducted using the dimensions of an actual Cardan joint which had input and output links with masses of 38.115 g and a coupler of mass 44.352 g. The mass in the input and output links was distributed so that the first moments of inertia were S_x=0 and S_y=S_z=457.38 g-mm and the second moments of inertia were J_{xx}=1794.581 g-mm^2, J_{yy}=7575.356 g-mm^2 and J_{zz}=8855.385 g-mm^2. The first moments of inertia of the coupler were S_x=0 and S_y=S_z=731.808 g-mm and the second moments of inertia were J_{xx}=16332.432 g-mm^2, J_{yy}= 17071.824 g-mm^2 and J_{zz}=2498.496 g-mm^2. The products of inertia for the three moving links were negligible.

The calculations were taken for a ground-link twist angle of 160° at an input torque of 100 Nm and an input speed of 3000 rpm. These values represent a reasonable application of the Cardan joint under conditions where inertia effects are important.

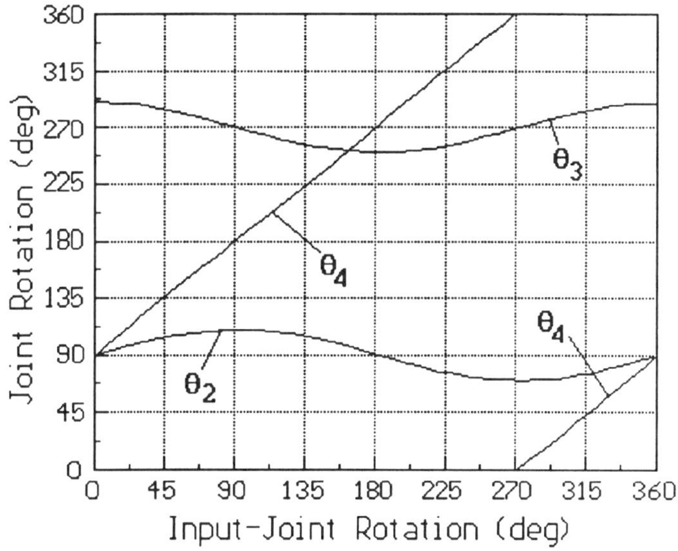

Figure 65 Joint rotational displacements of an ideal Cardan joint.

The computed joint rotational displacements are shown in Figure 65. The output rotation shows a slight double-frequency lead-and-lag fluctuation relative to the constant input rotation. The rotations at the intermediate joints are oscillatory, exhibiting single-frequency behavior.

The rotational speeds of the intermediate and output joints are shown in Figure 66. In this illustration it is clearly seen how the output shaft of a Cardan joint leads and lags the input shaft, averaging the input speed over the entire cycle. The oscillatory motion of the intermediate-joint rotational speeds is also apparent.

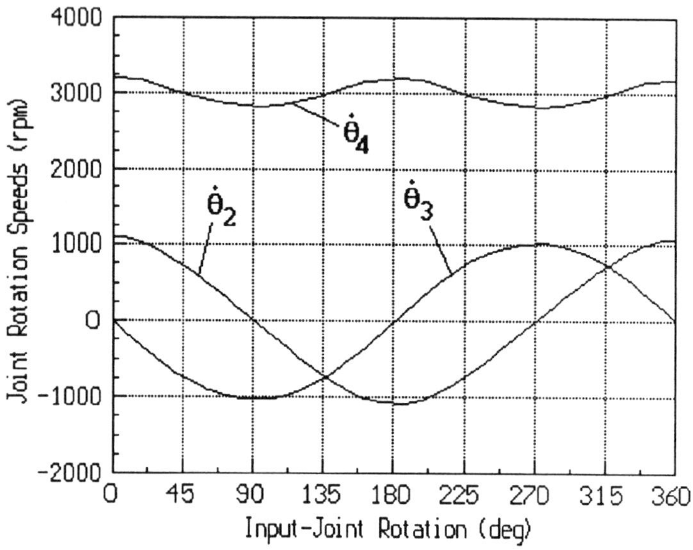

Figure 66 Joint rotational speeds of a Cardan joint.

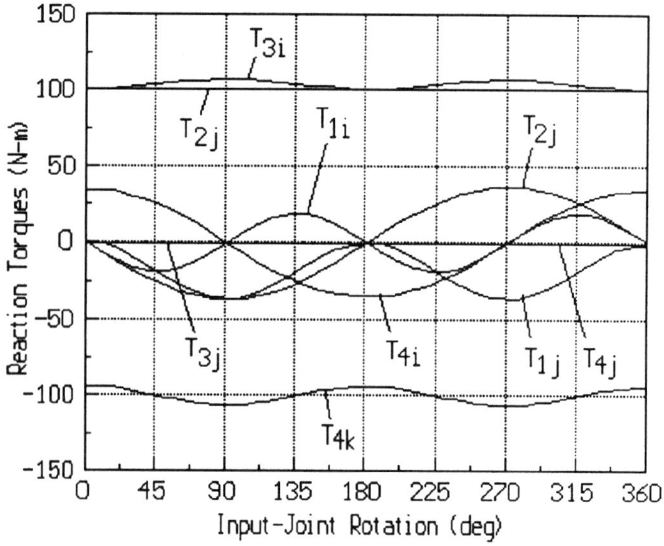

Figure 67 Reaction torques in a Cardan joint.

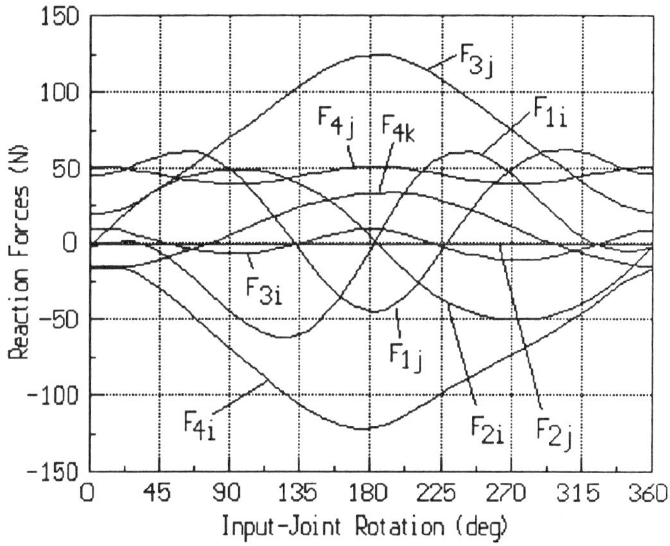

Figure 68 Reaction forces in a Cardan joint.

The reaction torques are shown in Figure 67. Torque components T_{2j} and T_{3j} are constant, equal in magnitude to the input torque. The torque component T_{4j} has negligible magnitude. The other torque components vary with either single or double frequency. The output torque, T_{4k} has a double frequency, averaging at a magnitude equal to the magnitude of the input torque and directed outward to load the machinery driven by the Cardan joint.

The variations in the reaction-force components are shown in Figure 68. Force components F_{2j} and F_{3j} are negligible, but the other force components acting on the joints can have significant magnitudes.

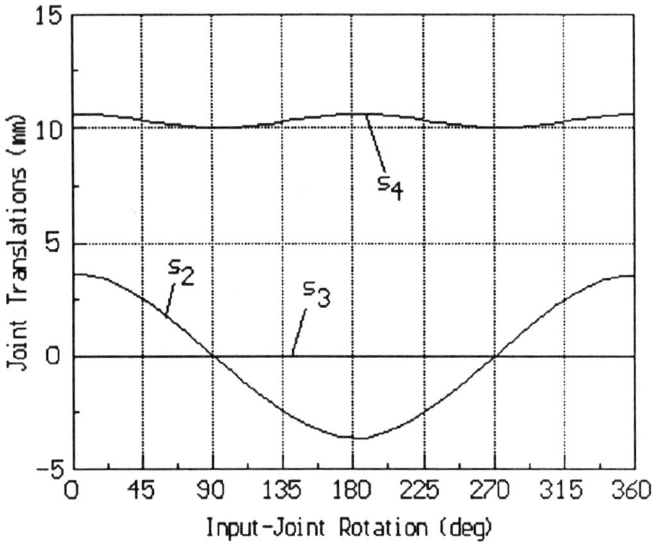

Figure 69 Translation in the joints of a pin-and-block Cardan joint.

The Pin-and-Block Cardan Joint

Because of manufacturing-tolerance and installation errors, a Cardan joint may not have ideal proportions. There is also a version of the Cardan joint where the joint axes of the coupler are purposely offset. This is called the "pin-and-block" Cardan joint, and its coupler is made of a rectangular length of rod featuring joints with axes located a significant distance from each other. While the pin-and-block Cardan joint offers economy of manufacture, the length of the coupler gives rise to translational motions which render it suitable only for low-speed applications. The translations for a Cardan joint with a coupler length a_2 = 10 mm and input translation $s_1 = 0$ are shown in Figure 69 where it is seen that there is no translation at joint 3 but there is a single-frequency translational displacement at the joint between the input crank and the coupler and a double-frequency translation at the output joint. The speeds of these translations are shown in Figure 70.

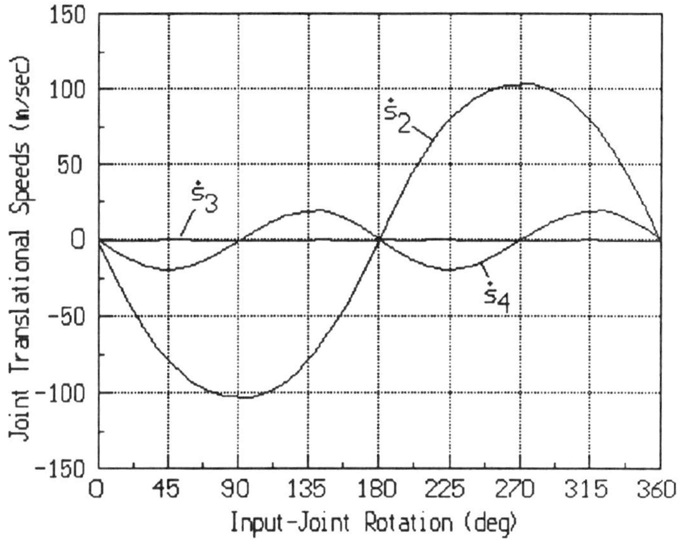

Figure 70 Translational speeds in a pin-and-block Cardan joint.

Statement List, Duplex.hpp

```
// source: duplex.hpp
// use:    Defines duplex arithmetic class
//          Implements inline functions
//          Based on complex.h by Capper, p.269 et al.
//          Ian S. Fischer
//          6/10/97

#ifndef DUPLEX_hpp
#define DUPLEX_hpp

#include <iostream.h>
#include <math.h>

class duplex {
    friend duplex operator+(double x, const duplex &v);
    friend duplex operator+(const duplex &u, double x);
    friend duplex operator+(const duplex &u, const duplex &v);

    friend duplex operator-(double x, const duplex &v);
    friend duplex operator-(const duplex &u, double x);
    friend duplex operator-(const duplex &u, const duplex &v);

    friend duplex operator*(double x, const duplex &v);
    friend duplex operator*(const duplex &u, double x);
    friend duplex operator*(const duplex &u, const duplex &v);
```

```
friend duplex operator/(const duplex &u, const duplex &v);
friend duplex operator/(const duplex &u, const double &v);
friend duplex operator/(const double &u, const duplex &v);

friend double prim(const duplex &z);
friend double dual(const duplex &z);

friend double mod(const duplex &z);
friend duplex conj(const duplex &z);

friend duplex sin(const duplex &z);
friend duplex cos(const duplex &z);
friend duplex tan(const duplex &z);

friend duplex deg2rad(const duplex &z);
friend double deg2rad(const double &z);
friend duplex rad2deg(const duplex &z);
friend double rad2deg(const double &z);

friend duplex rpm2radpersec(const duplex &z);
friend double rpm2radpersec(const double &z);
friend duplex radpersec2rpm(const duplex &z);
friend double radpersec2prm(const double &z);

friend duplex pow(const duplex &z, double y);
friend duplex sqrt(const duplex &z);

friend duplex atan(const duplex &z);
friend duplex atan2(const duplex &s, const duplex &c);

friend ostream &operator<<(ostream &os, const duplex &z);
friend istream &operator>>(istream &is, duplex &z);

public:
    double &prim(void);
    double &dual(void);

    duplex(void) { }
    duplex(double p, double d);
    duplex(const duplex &z);

    duplex &operator=(double x);
    duplex &operator=(const duplex &z);

    duplex &operator+=(double x);
    duplex &operator+=(const duplex &z);

    duplex operator-() const;

private:
    double pr, du;
};

// friend functions:

inline duplex operator+(double x, const duplex &v)
```

```
// Addition of a real and a duplex.
{
    return duplex(x + v.pr, v.du);
}

inline duplex operator+(const duplex &u, double x)
// Addition of a duplex and a real.
{
    return duplex(u.pr + x, u.du);
}

inline duplex operator+(const duplex &u, const duplex &v)
// Addition of a duplex and a duplex.
{
    return duplex(u.pr + v.pr, u.du + v.du);
}

inline duplex operator-(double x, const duplex &v)
// Subtraction of a duplex from a real.
{
    return duplex(x - v.pr, -v.du);
}

inline duplex operator-(const duplex &u, double x)
// Subtraction of a real from a duplex.
{
    return duplex(u.pr - x, u.du);
}

inline duplex operator-(const duplex &u, const duplex &v)
// Subtraction of a duplex from a duplex.
{
    return duplex(u.pr - v.pr, u.du - v.du);
}

inline duplex operator*(double x, const duplex &v)
// Multiplication of a real and a duplex.
{
    return duplex(x * v.pr, x * v.du);
}

inline duplex operator*(const duplex &u, double x)
// Multiplication of a duplex and a real.
{
    return duplex(u.pr * x, u.du * x);
}

inline duplex operator*(const duplex &u, const duplex &v)
// Multiplication of a duplex and a duplex.
{
    return duplex(u.pr * v.pr, u.pr * v.du + u.du * v.pr);
}

inline duplex operator/(const duplex &u, const duplex &v)
// Division of a duplex by a duplex.
{
    return duplex(u.pr/v.pr,
```

```
                (u.du * v.pr - u.pr * v.du) / ( v.pr * v.pr ) );
}

inline duplex operator/(const duplex &u, const double &v)
// Division of a duplex by a double.
{
    return duplex(u.pr/v, u.du/v);
}

inline duplex operator/(const double &u, const duplex &v)
// Division of a double by a duplex.
{
    return duplex( u/v.pr, - u * v.du / ( v.pr * v.pr ) );
}

inline double prim(const duplex &z)
// Return the primary part of a duplex.
{
    return z.pr;
}

inline double dual(const duplex &z)
// Return the dual part of a duplex.
{
    return z.du;
}

inline double mod(const duplex &z)
// modulus of z
{
    return(sqrt(z.pr * z.pr + z.du * z.du));
}

inline duplex conj(const duplex &z)
// Complex conjugate of z.
{
    return duplex(z.pr, -z.du);
}

inline duplex sin(const duplex &z)
// Sine function for duplex argument.
{
    return duplex(sin(z.pr), z.du * cos(z.pr));
}

inline duplex cos(const duplex &z)
// Cosine function for duplex argument.
{
    return duplex(cos(z.pr), -z.du * sin(z.pr));
}

inline duplex tan(const duplex &z)
// Tangent function for duplex argument.
{
    return duplex( tan(z.pr), z.du/( cos(z.pr)*cos(z.pr) ) );
}
```

```
inline duplex atan2(const duplex &s, const duplex &c)
// Double-argument tangent function for duplex arguments.
{
    double pi=4.*atan(1.);
    if (s.pr > 0 && c.pr > 0)
        return duplex( atan( s/c ) );
    if (c.pr < 0 )
        return duplex( atan( s/c ) + pi );
    else
        return duplex( atan( s/c ) + 2*pi );
}

inline duplex deg2rad(const duplex &z)
// Convert primary portion of duplex angle from degrees to radians.
{
    return duplex(z.pr*atan(1.)/45.,z.du);
}

inline double deg2rad(const double &z)
// Convert primary angle from degrees to radians.
{
    return double(z*atan(1.)/45.);
}

inline duplex rad2deg(const duplex &z)
// Convert primary portion of duplex angle from radians to degrees.
{
    double zdeg = z.pr*45./atan(1.); // convert angle to degrees
    while (zdeg < 0.) {
        zdeg = zdeg + 360.;          // adjust angle to positive
    }
    while (zdeg > 360.) {            // adjust angle to < 360 deg
        zdeg = zdeg - 360.;
    }
    return duplex(zdeg,z.du);
}

inline double rad2deg(const double &z)
// Convert primary angle from radians to degrees.
{
    double zdeg = z*45./atan(1.);    // convert angle to degrees
    while (zdeg < 0.) {
        zdeg = zdeg + 360.;          // adjust angle to positive
    }
    while (zdeg > 360.) {            // adjust angle to < 360 deg
        zdeg = zdeg - 360.;
    }
    return zdeg;
}

inline duplex rpm2radpersec(const duplex &z)
// Convert primary portion of duplex velocity from rpm to rad/sec.
{
    return duplex(z.pr*4.*atan(1.)/30.,z.du);
}

inline double rpm2radpersec(const double &z)
```

```
// Convert primary velocity from rpm to radians/second.
{
    return double(z*4.*atan(1.)/30.);
}

inline duplex radpersec2rpm(const duplex &z)
// Convert primary portion of duplex angle from radians to degrees.
{
    return duplex(z.pr*30./(4.*atan(1.)),z.du);
}

inline double radpersec2rpm(const double &z)
// Convert primary angle from radians to degrees.
{
    return double(z*30./(4.*atan(1.)));
}

inline duplex pow(const duplex &z, double y)
// Duplex number raised to a power.
{
    return duplex(pow(z.pr,y), y * z.du * pow(z.pr,y-1.));
}

inline duplex sqrt(const duplex &z)
// Square root of a duplex argument.
{
    return duplex(sqrt(z.pr), 0.5 * z.du / sqrt(z.pr));
}

inline duplex atan(const duplex &z)
// Arctangent of a duplex argument.
{
    return duplex(atan(z.pr), z.du/(1.+z.pr*z.pr));
}

// member functions and operators:

inline double &duplex::prim(void)
// primary component of a duplex number
{
    return pr;
}

inline double &duplex::dual(void)
// dual component of a duplex number
{
    return du;
}

inline duplex::duplex(double p, double d)
// constructor
{
    pr = p;
    du = d;
}

inline duplex::duplex(const duplex &z)
```

```
// copy constructor
{
    pr = z.pr;
    du = z.du;
}

inline duplex &duplex::operator=(double x)
// overload assignment operator
{
    pr = x;
    du = 0.0;
    return *this;
}

inline duplex &duplex::operator=(const duplex &z)
// overload assignment operator
{
    pr = z.pr;
    du = z.du;
    return *this;
}

inline duplex &duplex::operator+=(double x)
// overload increment operator
{
    pr += x;
    return *this;
}

inline duplex &duplex::operator+=(const duplex &z)
// overload increment operator
{
    pr += z.pr;
    du += z.du;
    return *this;
}

inline duplex duplex::operator-() const
// negative of a duplex number (unary minus)
{
    return duplex(-pr, -du);
}

ostream &operator<<(ostream &os, const duplex &z)
// output stream insertion operator
{
    os << '(' << z.pr << ',' << z.du << ')' ;
    return os ;
}

istream &operator>>(istream &is, duplex &z)
// input stream extraction operator
{
    is >> z.pr >> z.du;
    return is ;
}
```

```
#endif // DUPLEX_h
```

Statement List, RCCC.cpp

```
// RCCC.cpp
// Program to calculate
// displacements, joint speeds, absolute velocities,
// momenta, derivatives of momenta, inertia forces and reaction forces
// in a spatial four-bar mechanism
// Ian S. Fischer
// 6/11/97

#include <math.h>
#include <iostream.h>
#include <fstream.h>
#include "duplex.hpp"

ofstream fp;

class start_and_step{
public:
    double step, th1, s1, th1dot, s1dot;
    void get_start_and_step(void);
    void default_start_and_step(void);
};

class linkage {
public:
    duplex alpha[5], theta[5], vel[5], v[4][3],
      H[4][3], Hprev[4][3], Hdot[4][3], f[4][3], F[5][3];
    double m[4], sx[4], sy[4], sz[4],
      jxx[4], jyy[4], jzz[4], jxy[4], jxz[4], jyz[4];

    void get_links(void);
    void default_links(void);
    void get_input_torque_and_axial_forces();
    void default_input_torque_and_axial_forces();

    void displacements(double th1, double s1);
    void velocities(double th1dot, double s1dot);
    void absolute_velocities(void);
    void momenta(void);
    void derivatives_of_momenta(double th1dot, double step);
    void inertia_forces(void);
    void reaction_forces(void);

    void display_displacements(void);
    void display_velocities(void);
    void display_absolute_velocities(void);
    void display_momenta(void);
    void display_inertia_forces(void);
    void display_reaction_forces(void);

    void write_to_file(void);
};
```

```
void linkage::get_links(void)
{
    int n;

    for ( n = 1 ; n <= 4 ; n++ )
    {
        cout << "\nTwist and length of link " << n << " (2 items) ? ";
        cin >> alpha[n];
        alpha[n] = deg2rad( alpha[n] );

        if( n < 4 ) {
            cout << "Mass of link " << n << " ? ";
            cin >> m[n];

            cout << "First moments of inertia Sx, Sy, Sz of link "
                 << n <<" (3 items) ?\n";
            cin >> sx[n] >> sy[n] >> sz[n];

            cout << "Second moments of inertia Jxx, Jyy, Jzz of link "
                 << n << " (3 items) ?\n";
            cin >> jxx[n] >> jyy[n] >> jzz[n];

            cout << "Products of inertia Jxy, Jxz, Jyz of link "
                 << n << " (3 items) ?\n";
            cin >> jxy[n] >> jxz[n] >> jyz[n];
        }
    }
}

void linkage::default_links(void)
{
    alpha[1] = deg2rad(duplex(90., 0.));
    alpha[2] = deg2rad(duplex(90., 0.));
    alpha[3] = deg2rad(duplex(90., 0.));
    alpha[4] = deg2rad(duplex(160., 0.));

    m[1] = 0.038115;
    m[2] = 0.044352;
    m[3] = 0.038115;

    sx[1] = 0.;  sy[1] = 4.57380E-4; sz[1] = 4.57380E-4;
    sx[2] = 7.31808E-4; sy[2] = 7.31808E-4; sz[2] = 0.;
    sx[3] = 0.;  sy[3] = 4.57380E-4; sz[3] = 4.57380E-4;

    jxx[1] = 1.794581E-6; jyy[1] = 7.575356E-6; jzz[1] = 8.855385E-6;
    jxx[2] = 1.633243E-5; jyy[2] = 1.707182E-5; jzz[2] = 2.498496E-6;
    jxx[3] = 1.794581E-6; jyy[3] = 7.575356E-6; jzz[3] = 8.855385E-6;

    jxy[1] = 0.; jyz[1] = 0.; jxz[1] = 0.;
    jxy[2] = 0.; jyz[2] = 0.; jxz[2] = 0.;
    jxy[3] = 0.; jyz[3] = 0.; jxz[3] = 0.;
}

void start_and_step::get_start_and_step(void)
{
```

```
    cout << "\nInput translation? " ;
    cin >> s1;

    cout << "\nStep size of theta1? " ;
    cin >> step;

    th1 = -step;

    cout << "\nInput angular (RPM) and linear velocities " ;
    cin >> th1dot >> s1dot;
}

void start_and_step::default_start_and_step(void)
{
    s1 = 0.;
    step = 10.;
    th1 = -step;
    th1dot = 3000.;
    s1dot = 0.;
}

void linkage::get_input_torque_and_axial_forces(void)
{
    double T1k, F1k, F2k, F3k;

    cout << "\nInput torque? ";
    cin >> T1k;

    cout <<
       "\nAxial forces on input and intermediate joints (3 items)?\n";
    cin >> F1k >> F2k >> F3k;

    F[1][2] = duplex(F1k, T1k);
    F[2][2] = duplex(F2k, 0.);
    F[3][2] = duplex(F3k, 0.);
}

void linkage::default_input_torque_and_axial_forces(void)
{
    double T1k, F1k, F2k, F3k;

    T1k = 100.;

    F1k = 0.;
    F2k = 0.;
    F3k = 0.;

    F[1][2] = duplex(F1k, T1k);
    F[2][2] = duplex(F2k, 0.);
    F[3][2] = duplex(F3k, 0.);
}

void main()
{
    int n ;
    char choice, choice_file, output_file[80];
```

```
    linkage mechanism;
    start_and_step index;

    cout <<
           "RCCC\n(when multiple items requested, separate by spaces)\
    \n\nUse default mechanism (y/n)? ";
    cin >> choice;

    if (choice == 'y') {
        mechanism.default_links();
        index.default_start_and_step();
        mechanism.default_input_torque_and_axial_forces();
    }
    else {
        mechanism.get_links();
        index.get_start_and_step();
        mechanism.get_input_torque_and_axial_forces();
    }

    cout << "\nWrite results to disk file (y/n)? ";
    cin >> choice_file;

    if (choice_file == 'y') {
        cout << "\nName of output disk file? ";
        cin >> output_file;
        fp.open(output_file, ios::out);

     // First line of data file consists of the column headers

        fp << "th1 s1 th1dot s1dot F1i T1i F1j T1j F1k T1k "\
              "th2 s2 th2dot s2dot F2i T2i F2j T2j F2k T2k "\
              "th3 s3 th3dot s3dot F3i T3i F3j T3j F3k T3k "\
              "th4 s4 th4dot s4dot F4i T4i F4j T4j F4k T4k\n";
    }

    while (index.th1 <= 360)
    {
        mechanism.displacements(index.th1, index.s1);
        mechanism.velocities(index.th1dot, index.s1dot);
        mechanism.absolute_velocities();
        mechanism.momenta();

        mechanism.derivatives_of_momenta(index.th1dot, index.step);
        mechanism.inertia_forces();
        mechanism.reaction_forces();

        mechanism.display_displacements();
        mechanism.display_velocities();
        mechanism.display_absolute_velocities();
        mechanism.display_momenta();
        mechanism.display_inertia_forces();
        mechanism.display_reaction_forces();

        if (choice_file == 'y' && index.th1 >= 0) {
            mechanism.write_to_file();
        }
```

```
        index.th1 += index.step;
    }

    if (choice_file == 'y') {
        cout << "\nData written to file " << output_file << "\n";
        fp.close();
    }
}

void linkage::display_displacements(void)
{
        cout.precision(7);
        cout << "theta = " << rad2deg(theta[1])
            << " " << rad2deg(theta[2])
            << " " << rad2deg(theta[3])
            << " " << rad2deg(theta[4]) << "\n";
}

void linkage::display_velocities(void)
{
        cout.precision(7);
        cout << "speeds = " << vel[1] << " " << vel[2] <<
                        " " << vel[3] << " " << vel[4] << "\n";
}

void linkage::display_absolute_velocities(void)
{
    cout.precision(7);
    for (int n = 1; n <= 3; n++) {
        cout << "abs_vel" << n << " = " << v[n][0] << " "
             << v[n][1] << " " << v[n][2] << "\n";
    }
}

void linkage::display_momenta(void)
{
    cout.precision(7);
    for (int n = 1; n <= 3; n++) {
        cout << "H" << n << " = " << H[n][0] << " "
             << H[n][1] << " " << H[n][2] << "\n";
    }
    for (n = 1; n <= 3; n++) {
        cout << "Hprev" << n << " = " << Hprev[n][0] << " "
             << Hprev[n][1] << " " << Hprev[n][2] << "\n";
    }
    for (n = 1; n <= 3; n++) {
        cout << "Hdot" << n << " = " << Hdot[n][0] << " "
             << Hdot[n][1] << " " << Hdot[n][2] << "\n";
    }
}

void linkage::display_inertia_forces(void)
{
    for (int n = 1; n <= 3; n++) {
        cout << "f" << n << " = " << f[n][0] << " "
             << f[n][1] << " " << f[n][2] << "\n";
    }
```

```
}

void linkage::display_reaction_forces(void)
{
    for (int n = 1; n <= 4; n++) {
        cout << "F" << n << " = " << F[n][0] << " "
             << F[n][1] << " " << F[n][2] << "\n";
    }
}

void linkage::write_to_file(void)
{
    for (int n = 1; n <=4; n++) {// index through links n

        fp << prim(rad2deg(theta[n])) << " " << dual(theta[n]) << " "
           << prim(vel[n]) << " " << dual(vel[n]) << " ";

        for (int i = 0; i <= 2; i++) { // index through directions
                                          i,j,k
           fp << prim(F[n][i]) << " " << dual(F[n][i]) << " ";
        }

     }
     fp << "\n"; // line shift each time this function has been called
}



void linkage::displacements( double th1, double s1 )
{
    duplex D, E, F, t, st2, ct2, st3, ct3;

    theta[1] = deg2rad( duplex( th1, s1 ) ) ;

    // Calculation of output dual angle theta[4]

    D = sin(alpha[1]) * sin(alpha[3]) * sin(theta[1]) ;
    E = -sin(alpha[3])*(cos(alpha[1])*sin(alpha[4])
                +sin(alpha[1])*cos(alpha[4])*cos(theta[1])) ;
    F = cos(alpha[3])*(cos(alpha[1])*cos(alpha[4])
                -sin(alpha[1])*sin(alpha[4])*cos(theta[1]))
                -cos(alpha[2]);

    t = ( -D - sqrt ( D*D + E*E - F*F ) ) / ( F - E ) ;

    theta[4] = 2.* atan ( t );

    // Calculation of intermediate dual angle theta[2]

    st2 = ( ( cos(alpha[3]) * sin(alpha[4])
          + sin(alpha[3]) * cos(alpha[4]) * cos(theta[4]) )
          * sin(theta[1])
          + sin(alpha[3]) * cos(theta[1]) * sin(theta[4]) )
          / sin(alpha[2]);

    ct2 = ( cos(alpha[1]) * cos(alpha[2])
```

```
        - cos(alpha[3]) * cos(alpha[4])
        + sin(alpha[3]) * sin(alpha[4]) * cos(theta[4]) )
        / ( sin(alpha[1]) * sin(alpha[2]) );

    theta[2] = atan2( st2, ct2);

    // Calculation of intermediate dual angle theta[3]

    st3 = ( cos(alpha[1]) * sin(alpha[4]) * sin(theta[1])
          + sin(alpha[1]) *
          ( sin(theta[1]) * cos(theta[4])
          + cos(alpha[4]) * cos(theta[1]) * sin(theta[4]) ) )
          / sin(alpha[2]);

    ct3 = ( - cos(alpha[1]) * cos(alpha[4])
          + cos(alpha[2]) * cos(alpha[3])
          + sin(alpha[1]) * sin(alpha[4]) * cos(theta[1]) )
          / ( sin(alpha[2]) * sin(alpha[3]) );

    theta[3] = atan2( st3, ct3 );
}

void linkage::velocities( double th1dot, double s1dot )
{
    vel[1] = duplex(th1dot, s1dot);

    vel[2] = - ((( cos(alpha[1]) * sin(alpha[2])
                 + sin(alpha[1]) * cos(alpha[2]) * cos(theta[2])
                 ) * sin(theta[3])
                + sin(alpha[1]) * sin(theta[2]) * cos(theta[3]) )
                / (sin(alpha[2]) * sin(theta[3]))
               ) * vel[1];

    vel[3] = ((( sin(alpha[2]) * cos(alpha[3])
                 + cos(alpha[2]) * sin(alpha[3]) * cos(theta[3])
               ) * sin(theta[2])
               + sin(alpha[3]) * cos(theta[2]) * sin(theta[3]) )
              / (sin(alpha[2]) * sin(alpha[3]) * sin(theta[3]))
             ) * sin(alpha[1]) * vel[1];

    vel[4] = - ( (sin(alpha[1]) * sin(theta[2]))
                 /(sin(alpha[3]) * sin(theta[3]))
               ) * vel[1];
}

void linkage::absolute_velocities( void )
{
    int i = 0, j = 1, k = 2;

    duplex v1 = rpm2radpersec(vel[1]);
    duplex v3 = rpm2radpersec(vel[3]);
    duplex v4 = rpm2radpersec(vel[4]);

    v[1][i] = 0.;
    v[1][j] = v1 * sin(alpha[1]);
    v[1][k] = v1 * cos(alpha[1]);
```

```
    v[2][i] = -v4 * sin(alpha[3]) * sin(theta[3]);
    v[2][j] =  v4 * sin(alpha[3]) * cos(theta[3]);
    v[2][k] = -v4 * cos(alpha[3]) - v3;

    v[3][i] = 0.;
    v[3][j] = 0.;
    v[3][k] = -v4;
}

void linkage::momenta( void)
{
    int n, i = 0, j = 1, k = 2;

    // Save present value of momenta for use in calculation of
    // derivatives of momenta
    for( n = 1; n <= 3; n++) {
        Hprev[n][i] = H[n][i];
        Hprev[n][j] = H[n][j];
        Hprev[n][k] = H[n][k];
    }

    // Calculation of dual momenta
    for( n = 1; n <= 3; n++) {
        H[n][i] = duplex (
                    m[n] * dual(v[n][i])
                    + sz[n] * prim(v[n][j]) - sy[n] * prim(v[n][k]),
                    - sz[n] * dual(v[n][j]) + sy[n] * dual(v[n][k])
                    + jxx[n] * prim(v[n][i])
                    - jxy[n] * prim(v[n][j]) - jxz[n] * prim(v[n][k]) );
        H[n][j] = duplex (
                    m[n] * dual(v[n][j])
                    - sz[n] * prim(v[n][i]) + sx[n] * prim(v[n][k]),
                    + sz[n] * dual(v[n][i]) - sx[n] * dual(v[n][k])
                    + jyy[n] * prim(v[n][j])
                    - jxy[n] * prim(v[n][i]) - jyz[n] * prim(v[n][k]) );
        H[n][k] = duplex (
                    m[n] * dual(v[n][k])
                    + sy[n] * prim(v[n][i]) - sx[n] * prim(v[n][j]),
                    - sy[n] * dual(v[n][i]) + sx[n] * dual(v[n][j])
                    + jzz[n] * prim(v[n][k])
                    - jxz[n] * prim(v[n][i]) - jyz[n] * prim(v[n][j]) );

    }
}

void linkage::derivatives_of_momenta( double th1dot, double step )
{
    int n, i;
    for ( n =1 ; n <= 3; n++) {
        for (i = 0; i <=2; i++) {
            Hdot[n][i] = (H[n][i] - Hprev[n][i])
                            * ( 6. * th1dot/ step );
            // factor of 6 converts th1dot from rpm to deg/sec
        }
    }
}
```

```
void linkage::inertia_forces(void)
{
    int i = 0, j = 1, k = 2;
    for ( int n = 1; n <= 3; n++) {
        f[n][i] = Hdot[n][i] - v[n][k] * H[n][j] + v[n][j] * H[n][k];
        f[n][j] = Hdot[n][j] - v[n][i] * H[n][k] + v[n][k] * H[n][i];
        f[n][k] = Hdot[n][k] - v[n][j] * H[n][i] + v[n][i] * H[n][j];
    }
}

void linkage::reaction_forces()
{
    duplex K[10];
    int i = 0, j = 1, k = 2;

    K[1] = f[1][i];

    K[2] = f[1][j] - F[1][k] * sin(alpha[1]);

    K[3] = f[1][k] - F[1][k] * cos(alpha[1]) + F[2][k];

    K[4] = f[2][i];

    K[5] = f[2][j] - F[2][k] * sin(alpha[2]);

    K[6] = f[2][k] - F[2][k] * cos(alpha[2]) + F[3][k];

    K[7] = f[3][i];

    K[8] = f[3][j] - F[3][k] * sin(alpha[3]);

    K[9] = f[3][k] - F[3][k] * cos(alpha[3]);

    F[1][i] = (
                 K[1] * sin(alpha[1]) * sin(alpha[2])
                    * cos(theta[1]) * sin(theta[2])
                -K[2] * sin(alpha[1]) * sin(alpha[2])
                    * cos(theta[1]) * cos(theta[2])
                +K[3] * sin(alpha[2])
                    * ( sin(theta[1]) * sin(theta[2])
                      - cos(alpha[1]) * cos(theta[1]) * cos(theta[2])
)
                +K[6] * sin(alpha[1]) * cos(theta[1])
               )
                /( sin(alpha[1]) * sin(alpha[2]) * sin(theta[2]) );

    F[1][j] = (
                 K[1] * sin(alpha[1]) * sin(alpha[2])
                    * sin(theta[1]) * sin(theta[2])
                -K[2] * sin(alpha[1]) * sin(alpha[2])
                    * sin(theta[1]) * cos(theta[2])
                -K[3] * sin(alpha[2])
                    * ( cos(theta[1]) * sin(theta[2])
                      + cos(alpha[1]) * sin(theta[1]) * cos(theta[2])
)
                +K[6] * sin(alpha[1]) * sin(theta[1])
               )
```

```
            /( sin(alpha[1]) * sin(alpha[2]) * sin(theta[2]) );

F[2][i] = (
              -K[2] * sin(alpha[1]) * sin(alpha[2]) * cos(theta[2])
              -K[3] * cos(alpha[1]) * sin(alpha[2]) * cos(theta[2])
              +K[6] * sin(alpha[1])
            )
            / ( sin(alpha[1]) * sin(alpha[2]) * sin(theta[2]) );

F[2][j] = -K[3] * cos(alpha[1]) / sin(alpha[1]) - K[2];

F[3][i] = (
              -K[2] * sin(alpha[1]) * sin(alpha[2])
              -K[3] * cos(alpha[1]) * sin(alpha[2])
              -K[4] * sin(alpha[1]) * sin(alpha[2]) * sin(theta[2])
              +K[6] * sin(alpha[1]) * cos(theta[2])
            )
            / ( sin(alpha[1]) * sin(alpha[2]) * sin(theta[2]) );

F[3][j] = - K[6] * cos(alpha[2]) / sin(alpha[2]) - K[5];

F[4][i] = (
              - K[2] * sin(alpha[1]) * sin(alpha[2]) * cos(theta[3])
              - K[3] * cos(alpha[1]) * sin(alpha[2]) * cos(theta[3])
              - K[4] * sin(alpha[1]) * sin(alpha[2])
                     * sin(theta[2]) * cos(theta[3])
              - K[5] * sin(alpha[1]) * sin(alpha[2])
                     * sin(theta[2]) * sin(theta[3])
              + K[6] * sin(alpha[1])
                     * ( cos(theta[2]) * cos(theta[3])
                   - cos(alpha[2]) * sin(theta[2]) * sin(theta[3]) )
              - K[7] * sin(alpha[1]) * sin(alpha[2]) * sin(theta[2])
            )
            / ( sin(alpha[1]) * sin(alpha[2]) * sin(theta[2]) );

F[4][j] = (
               K[2] * sin(alpha[1]) * sin(alpha[2]) * cos(alpha[3])
                    * sin(theta[3])
             + K[3] * cos(alpha[1]) * sin(alpha[2]) * cos(alpha[3])
                    * sin(theta[3])
             + K[4] * sin(alpha[1]) * sin(alpha[2]) * cos(alpha[3])
                    * sin(theta[2]) * sin(theta[3])
             - K[5] * sin(alpha[1]) * sin(alpha[2]) * cos(alpha[3])
                    * sin(theta[2]) * cos(theta[3])
             - K[6] * sin(alpha[1]) * cos(alpha[3])
                    * (             cos(theta[2]) * sin(theta[3])
                 + cos(alpha[2]) * sin(theta[2]) * cos(theta[3]) )
             -K[8] * sin(alpha[1]) * sin(alpha[2]) * sin(theta[2])
            )
            / ( sin(alpha[1]) * sin(alpha[2]) * sin(theta[2]) );

F[4][k] = (
              - K[2] * sin(alpha[1]) * sin(alpha[2]) * sin(theta[3])
              - K[3] * cos(alpha[1]) * sin(alpha[2])
                     * sin(alpha[3]) * sin(theta[3])
              - K[4] * sin(alpha[2]) * sin(alpha[3]) * sin(alpha[3])
                     * sin(theta[2]) * sin(theta[3])
```

```
                + K[5] * sin(alpha[1]) * sin(alpha[2]) * sin(alpha[3])
                       * sin(theta[2]) * cos(theta[3])
                + K[6] * sin(alpha[1]) * sin(alpha[3])
                       * ( cos(theta[2]) * sin(theta[3])
                     + cos(alpha[2]) * sin(theta[2]) * cos(theta[3]) )
                - K[9] * sin(alpha[1]) * sin(alpha[2]) * sin(theta[2])
              )
              / ( sin(alpha[1]) * sin(alpha[2]) * sin(theta[2]) );

}
```

Bibliography

S.K. Agrawal "Multibody Dynamics: A Formulation Using Kane's Method and Dual Vectors" ASM *Journal of Mechanical Design* December 1993 (v.115) pp.833-838.

C. Bagci "Dynamic Force and Torque Analysis of Mechanisms Using Dual Vectors and 3×3 Matrix" ASME *Journal of Engineering for Industry* May 1972 (v.94) pp.738-745.

C. Bagci "Static Force and Torque Analysis Using 3×3 Screw Matrix, and Transmission Criteria for Space Mechanisms" ASME *Journal of Engineering for Industry* February 1971 (v.93) pp.90-101.

Louis Brand *Vector and Tensor Analysis* John Wiley & Sons, Inc., New York, 1947, pp.63-83.

C.K. Chen and F. Freudenstein "Dynamic Analysis of a Universal Joint With Manufacturing Tolerances" ASME *Journal of Mechanisms, Transmissions & Automation in Design* December 1986 (v.108) pp.524-532.

John J. Craig *Introduction to Robotics, Mechanics & Control* Addison-Wesley Publishing Co., Reading, MA, 1986, ISBN 0-201-10326-5.

I.S. Fischer "A Complete Notation for Dual Velocity" ASME *Journal of Mechanisms, Transmissions & Automation in Design* June 1992 (v.114) pp.263-268.

I.S. Fischer "Application of the Principle of Transference to the Evaluation of Translational Displacements in Spatial Mechanisms" *Trends and Developments in Mechanisms, Machines and Robotics - 1988*, v.2, pp.7-11 ASME DE-vol.15-2.

Ian S. Fischer "Application of a Generalized Wobble Plate Mechanism" Section 12.6, *Modern Kinematics, Developments in the Last Forty Years* (Arthur G. Erdman, ed.) John Wiley & Sons, Inc., New York, 1993, ISBN 0-471-55459-6, pp.564-571.

I.S. Fischer "Cardan-Joint Dynamics With Improved Assumptions" *Second National Applied Mechanisms & Robotics Conference*, 1991, v.2 paper VIIIB.3.

I.S. Fischer "Displacement Errors in Cardan Joints Caused by Coupler-Link Joint-Axes Offset" *First National Applied Mechanisms & Robotics Conference*, 1989, v.2, paper 89AMR-9A-2.

Ian S. Fischer "Dual Angle and Axis of a Screw Motion" *Mechanisms and Machine Theory* April 1998 (v.33) pp.331-340.

Ian S. Fischer "Dynamics of the Ideal Wobble-Plate Mechanism" *Journal of Applied Mechanisms & Robotics* July 1995 (v.2) pp.1-6

Ian S. Fischer "Inertia Effects in Cardan-Joint Operation" Proceedings of the 1996 *ASME Design Engineering Technical Conferences and Computers in Engineering Conference*, paper 96-DETC/MECH-1216.

Ian S. Fischer "Kinematics of the R7 Constant-Velocity Coupling" *Fifth National Applied Mechanisms and Robotics Conference*, October 13, 1997, paper AMR97-006.

Ian S. Fischer "Novel Methods in the Displacement Analysis of Spatial Mechanisms" *Mechanism Synthesis and Analysis*, 1994, pp.1-6, ASME DE-vol.70.

Ian S. Fischer and Johannes K. de Waal "Experimental Study of Cardan-Joint Dynamics" ASME *Journal of Mechanical Design* December 1995 (v.117) pp.526-531.

I.S. Fischer and F. Freudenstein "Internal Force and Moment Transmission in a Cardan Joint with Manufacturing Tolerances" ASME *Journal of Mechanisms, Transmissions and Automation in Design*, September 1984 (v.106) pp.301-311.

I.S. Fischer and R.N. Paul "Kinematic Displacements in a Double-Cardan-Joint Drive Shaft" ASME *Journal of Mechanical Design* September 1991 (v.113) pp.263-271.

I.S. Fischer and S. Rahman "Dynamics of the Generalized Slider-Crank Mechanism" *Advances in Design Automation - 1993*, v.1, ASME DE-vol.65-1 pp.333-340.

I.S. Fischer and S. Rahman "Kinematics of the Generalized Slider-Crank Mechanism" *Advances in Design Automation - 1993*, v.1, ASME DE-vol.65-1 pp.325-331.

I.S. Fischer and P.M. Remington "Errors in Constant-Velocity Shaft Couplings" ASME *Journal of Mechanical Design* March 1994 (v.116) pp.204-209.

L.M. Hsia and A.T. Yang "On the Principle of Transference in Three-Dimensional Kinematics" ASME *Journal of Mechanical Design* July 1981 (v.103) pp.652-656.

M.L. Keler "Kinematics and Statics Including Friction in Single-Loop Mechanisms by Screw Calculus and Dual Vectors" ASME *Journal of Engineering for Industry* May 1973 (v.95) pp.471-480.

J.M. McCarthy "Dual Orthogonal Matrices in Manipulator Kinematics" *International Journal of Robotics Research* Summer 1986 (v.5) pp.45-51.

J.M. McCarthy "On the Scalar and Dual Formulations of the Curvature Theory of Line Trajectories" ASME *Journal of Mechanisms, Transmissions & Automation in Design* March 1987 (v.109) pp.101-106.

J. Michael McCarthy *An Introduction to Theoretical Kinematics* The MIT Press, Cambridge, MA, 1990, ISBN 0-262-13242-4.

Richard P. Paul *Robot Manipulators, Mathematics, Programming and Control* The MIT Press, Cambridge, MA, 1981, ISBN 0-262-16082-X.

G.R. Pennock "Dynamic Analysis of Two Cooperating Robots" *Advances in Design Automation - 1987* v.2 pp.63-72 ASME DE-vol.10-2.

G.R. Pennock and B.A. Oncu "Application of Screw Theory to Rigid Body Dynamics" ASME *Journal of Dynamic Systems & Control* June 1992 (v.114) pp.262-269.

G.R. Pennock and A.T.Yang "Application of Dual-Number Matrices to the Inverse Kinematics Problem of Robot Manipulators" ASME *Journal of Mechanisms, Transmissions & Automation in Design* June 1985 (v.107) pp.201-208.

G.R. Pennock and A.T. Yang "Dynamic Analysis of a Multi-Rigid-Body Open-Chain System" ASME *Journal of Mechanisms, Transmissions & Automation in Design* March 1983 (v.105) pp.28-34.

G.R. Pennock and A.T. Yang "Instantaneous Kinematics of Three-Parameter Motions" ASME *Journal of Mechanisms, Transmissions & Automation in Design* June 1985 (v.107) pp.157-162.

B. Roth "On the Screw Axes and Other Special Lines Associated With Spatial Displacements of a Rigid Body" ASME *Journal of Engineering for Industry* February 1967 (v.89) pp.102-109.

B. Roth "Screws, Motors, and Wrenches That Cannot Be Bought in a Hardware Store" *Proceedings of the International Symposium on Robotics Research* August-September 1983 pp.679-694.

George N. Sandor and Arthur G. Erdman *Advanced Mechanism Design: Analysis and Synthesis* v.2 Prentice-Hall, Inc., Englewood Cliffs, NJ, 1984, ISBN 0-13-011437-5, p.621.

A.I. Sivridag, R. Gunes and S. Keles "The Serret-Frenet Formulae for a Dual Quaternion-Valued Functions of a Single Real Variable" *Mechanisms & Machine Theory* July 1994 (v.29) pp.749-754.

K. Sugimoto "Determination of Joint Velocities of Robots by Using Screws" ASME *Journal of Mechanisms, Transmissions & Automation in Design* June 1984 (v.106) pp.222-227.

J.J. Uicker, J. Denavit and R.S. Hartenberg "An Iterative Method for the Displacement Analysis of Spatial Mechanisms" *ASME Journal of Applied Mechanics* June 1964 (v.31) pp.309-314.

L.S. Woo and F. Freudenstein "Dynamic Analysis of Mechanisms Using Screw Coordinates" ASME *Journal of Engineering for Industry* February 1971 (v.93) pp.273-276.

A.T. Yang "Acceleration Analysis of Spatial Four-Link Mechanisms" ASME *Journal of Engineering for Industry* August 1966 (v.88) pp.296-300.

A.T. Yang "Analysis of an Offset Unsymmetric Gyroscope With Oblique Rotor Using (3×3) Matrices With Dual-Number Elements" ASME *Journal of Engineering for Industry* August 1969 (v.91) pp.535-542.

A.T. Yang "Application of Dual Quaternions to the Study of Gyrodynamics" ASME *Journal of Engineering for Industry* February 1967 (v.89) pp.137-143.

A.T. Yang "Calculus of Screws" *Basic Questions of Design Theory* (W.R. Spillers, ed.) American Elsevier Publishing Co. Inc., New York, 1974, ISBN 0444107398, pp.265-281.

A.T. Yang "Displacement Analysis of Spatial Five-Link Mechanisms Using (3×3) Matrices With Dual-Number Elements" ASME *Journal of Engineering for Industry* February 1969 (v.91) pp.152-157.

A.T. Yang "Inertia Force Analysis of Spatial Mechanisms" ASME *Journal of Engineering for Industry* February 1971 (v.93) pp.27-33.

A.T. Yang "Static Force and Torque Analysis of Spherical Four-Bar Mechanisms" ASME *Journal of Engineering for Industry* May 1965 (v.87) pp.221-227.

A.T. Yang and F. Freudenstein "Application of Dual-Number Quaternion Algebra to the Analysis of Spatial Mechanisms" ASME *Journal of Applied Mechanics* June 1964 (v.86) pp.300-308.

A.T. Yang, Y. Kirson and B. Roth "On a Kinematic Curvature Theory for Ruled Surfaces" *Fourth World Congress on the Theory of Machines and Mechanisms* Mechanical Engineering Publications Ltd., London, 1975.

Ziya Yapar "A Spatial Motion for a Dual Closed Strip" *Mechanism & Machine Theory* October 1994 (v.29) pp.1033-1042.

M.S.C. Yuan, F. Freudenstein and L.S. Woo "Kinematic Analysis of Spatial Mechanisms by Means of Screw Coordinates. Part 1 - Screw Coordinates" ASME *Journal of Engineering for Industry* February 1971 (v.93) pp.61-66.

M.S.C. Yuan, F. Freudenstein and L.S. Woo "Kinematic Analysis of Spatial Mechanisms by Means of Screw Coordinates. Part 2 - Analysis of Spatial Mechanisms" ASME *Journal of Engineering for Industry* February 1971 (v.93) pp.67-73.

Index